Industrial Maintenance

Industrial Maintenance

Terry Wireman

Reston Publishing Company, Inc.
A Prentice-Hall Company
Reston, Virginia

Library of Congress Cataloging in Publication Data

Wireman, Terry.
    Industrial maintenance.

    Includes index.
    1. Industrial equipment—Maintenance and repair.
I. Title.
TS191.W57 1983   621.8′16   82-24071
ISBN 0-8359-3070-X

Printed in the United States of America

Dedicated to my wife, Pam, and to my sons, Justin and Chad.

Contents

Preface

Inflation, design costs, labor costs, and other expenses have driven industrial equipment replacement costs skyward. It is becoming increasingly difficult for industrial corporations to replace old equipment with new equipment. As equipment becomes older it becomes necessary to have qualified and trained personnel to maintain and repair it. This group of people can greatly reduce the repair costs and increase productivity for any industrial facility. Competent and qualified repairmen know and understand the proper maintenance procedure, which enables the equipment to run at a high rate of efficiency and have a longer service life.

The problem arises in finding good qualified repairmen. It takes years of on-the-job experience working with competent people for the development of good repairmen. This time can be shortened by teaching the apprentice repairman the concepts behind his work, and instructing him in the whys and hows of what he is doing. This enables him to see why certain practices are followed, and why certain ones are to be avoided. Also, the journeyman repairman will need constant upgrading in the new methods and products on the market.

This book serves both of those purposes: it is designed to help the new apprentice from the first day on the job, working with the handtool until he progresses to the rank of repairman, and it will help the journeyman increase his productivity. This doesn't mean he has to work harder, just smarter.

The book covers all areas of industrial maintenance and repair. The more detailed theory is omitted because the book is written for repairmen, not engineers. Repairmen do not design the equipment but rather repair and maintain the equipment. This publication will benefit all repairmen and apprentices, but the benefit will be greatest for the industrial facility they work for. The guidelines will repay them with longer equipment life and less machine down time, which means increased productivity.

Chapter 1

Hand Tools

This chapter gives a description of the tools and their uses, and the care of the tools. In considering the hand tools of the repairman, let's consider the most prominent tools first.

HAMMERS

Ball Peen

Most repairmen are familiar with the many types and styles of hammers. The ball peen hammer is probably the most popular. It ranges in size from 4 oz to 2 lbs, the size chosen depending on the job to be done. The ball peen hammer head is shown in Figure 1-1 with the various parts named.

The hammer handle, usually wood, is inserted into the eye of the hammer and a small wedge is driven into the end of the hammer, spreading the wood out against the inside of the hammer eye. This wedging effect holds the hammer head on securely. If the head becomes loose the hammer should not be used until it is repaired by driving a new wedge into the wood in the eye of the hammer. If a repairman constantly uses his hammer, it'll develop burrs, or slivers, on the side of the face. If the hammer isn't repaired, sections of metal will break off the side of the

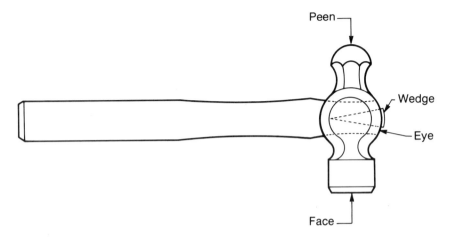

Figure 1-1. Ball peen hammer.

face, which can injure the repairman. It's best to grind off the burrs on the face while they are small. Excessive pressure shouldn't be used when grinding on the hammer. If the face becomes too hot it'll turn blue, which indicates the temper has been drawn from the face, leaving it soft and unfit for use.

Sledge Hammer

In addition to ball peen hammers, the repairman will on occasion use a sledge hammer. It's usually a rectangular shaped block of hardened steel with a wooden handle, used for much heavier applications. The most popular sizes are in the 8 to 12 lb range. The maximum force can be applied by swinging the hammer over the shoulder, using both arms, and striking the object being driven. Less force can be obtained by choking the hammer, that is, holding the hammer closer to the head and making short blows on the object being driven.

Soft Hammers

In addition to the sledge hammer the repairman will need a hammer made of a softer material. This is required so that when the worker hammers on hard finished material, the material won't be dented or damaged. The three most common are the rawhide hammer, plastic tip hammer, and the brass hammer (Figure 1-2). The brass hammer is the favorite of most

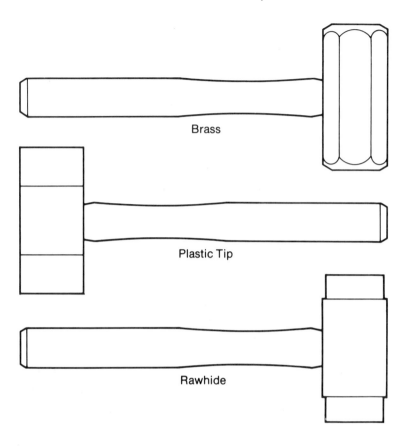

Brass

Plastic Tip

Rawhide

Figure 1-2. Soft hammers.

repairmen. The same care is given to the soft hammer as to the others. Remember that soft hammers wear much faster than hard hammers.

SCREWDRIVERS

The screwdriver is a vital part of any repairman's tool box. The screwdriver comes in a variety of sizes and styles. The *common, regular duty* screwdriver (Figure 1-3) has a handle with a steel rod embedded in the handle extending outward to the tip where the rod is squared and flattened out. The *heavy duty* screwdriver (Figure 1-4) has a square shank tapering into a tip at the end. The square shank is used to apply more torque to the screw by

Figure 1-3. Regular duty screwdriver.

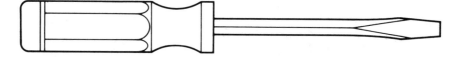

Figure 1-4. Heavy duty screwdriver.

turning it with a wrench, adjusted to the proper size, placed on the shank.

The other type of common screwdriver is the *phillips head* screwdriver which has a four-flute style tip. It is used exclusively with phillips head screws.

All of these screwdrivers are available in an almost limitless number of sizes. That is probably the most important consideration when selecting a screwdriver to use. Using an improper size screwdriver will damage the screw slot or the screwdriver. The tip of the screwdriver should fit snugly in the slot and should be about the same width as the screw slot. This enables the repairman to apply maximum force to the screwdriver without damage to the screwdriver tip or screw slot.

The three styles of screwdrivers are also available with many gimmicks that can save the repairman much time and frustration. These can range from magnetized screwdrivers to screwdrivers with clips to hold screws for starting in hard to reach areas. Offset screwdrivers are used when a conventional screwdriver cannot be used to reach the screw slot.

Screwdrivers are probably the most abused tool in the repairman's tool box. Most are used for anything from chisels to pry bars. Hammering on the handle of a screwdriver will mushroom the handle, drive the rod up into the handle, or break the tip. Prying with a screwdriver will bend the handle or break the tip. Either abuse will send the repairman to the store to buy a new screwdriver. If you break a tip on a screwdriver and would like to try to repair it, here is a method to try: Using a good, smooth grinding wheel, grind the tip square again, being careful not to get it too hot. (If it turns blue, you have overheated it and it'll

be too soft to continue using it.) Carefully grind the two sides almost parallel, then do the same to the faces. This will not only save the life of the screwdriver but also save money in screwdriver expense.

WRENCHES

Open End

There are many types of hand tools that fit into the wrench section. The first type considered are the open end wrenches (Figure 1-5). These are solid pieces of metal with U-shaped openings on each end, each the size of a particular nut or bolt head. The openings are slightly larger than the stamped size for clearance in putting them on the nuts or bolt heads. Open end wrenches have the heads set at an angle, which provides greater ease in assembling and disassembling parts in close quarters.

Box End

A wrench related to the open end is the box end wrench (Fig. 1-6), which completely encircles the nut or bolt head. These wrenches also come in a large variety of sizes, with clearance

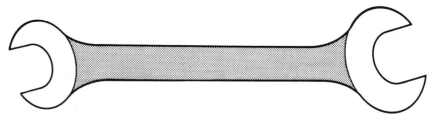

Figure 1-5. Open end wrench.

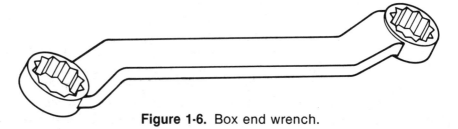

Figure 1-6. Box end wrench.

built in for aid in slipping the wrench over the nut or bolt head. The box end has notches cut into the circular part of the wrench for grasping the nut or bolt head. The number of notches corresponds to the point number of the wrench: six notches—six-point wrench; twelve notches—twelve-point wrench, for example. The head of the box wrench is usually offset from the handle of the wrench at a vertical angle, which provides clearance for the repairman's hands.

Heavier box end wrenches with a larger area of metal on one end designed to hit with a hammer are called *slugger* wrenches. This wrench is of special construction and is made of heavier material. No other box end wrench should be struck with a hammer, or have an extension put on it for more leverage. Standard wrenches are not designed for that type of torque input and will bend or break.

Combination

The combination wrench (Fig. 1-7) has an open end wrench on one end and a box end on the other. The repairman uses the box end to break nuts loose and the open for faster assembly and disassembly. Combination wrenches are also available in a large variety of sizes.

Socket Wrench

The socket set is a variation of the fixed wrenches, where a socket completely encloses the nut or bolt head. A ratchet or breaker bar is then inserted into the top of the socket to provide the lever arm needed to apply the proper torque. When in use, the ratchet (Fig. 1-8) applies torque in one direction while ratcheting in the other. This enables the repairman to work in close areas at a faster speed. The ratchet assembly is used for quick assembly and with a flip of a lever can be used for disassembly.

Socket sets include a number of accessories that make work easier. A breaker bar is provided, which is used for breaking loose very tight bolts or applying maximum torque when tightening. A speed handle is sometimes provided which gives maximum speed in assembling a large number of fasteners. A universal is also provided which enables a repairman to work in hard-to-reach areas. Socket extensions are also available for

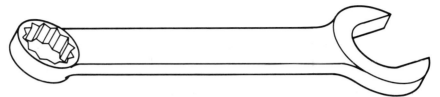

Figure 1-7. Combination wrench.

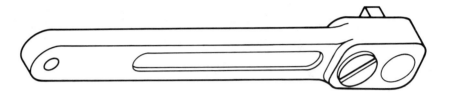

Figure 1-8. Ratchet.

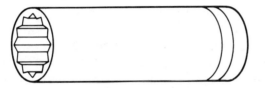 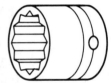

Figure 1-9. Socket wrench.

working in difficult-to-reach areas. There are a variety of sockets available to aid repairmen in working in difficult areas, from thin walled sockets with thin walls for working in tight places, to deep sockets (Fig. 1-9) for tightening nuts when the nut runs down on the bolt more than the thickness of the nut. The socket set, when properly cared for, will provide a lifetime of good service for the repairman.

Adjustable Wrench

The adjustable wrench (Fig. 1-10) is a very useful tool for the repairman. As the name implies it's a wrench that can adjust to many sizes. Usually the 8-, 10-, and 12-inch are the most popular sizes. This eliminates the need to carry a large number of combination wrenches. The wrench is shaped like an open end wrench, with one side of the opening fixed and the other side

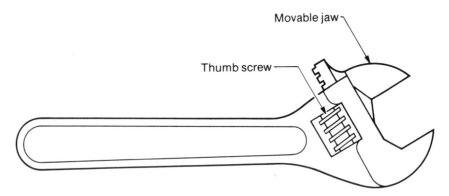

Figure 1-10. Adjustable wrench.

movable. It's opened wide enough to place it around the nut or bolt head; then the movable jaw can be tightened against the work. The wrench should be placed so that the fixed side of the jaw rather than the adjustable side is always pulled against. The movable jaw should be adjusted tightly against the work so that there is no chance of the wrench slipping.

Many repairmen misuse this tool, frequently hammering or pounding with it. These actions burr the end of the movable jaw of the wrench, preventing free movement of the jaw. When this does occur, the wrench should be disassembled, the jaw should be filed till smooth, and the burrs should be removed from the wrench. Once this is accomplished the wrench can be reassembled. Light oil should be applied to the adjustment screw and holding screw. The wrench should then be ready for use.

Pipe Wrenches

The pipe wrench (Fig. 1-11) is another type of adjustable wrench but it isn't used for bolts or nuts. Rather it's meant to be used on cylindrical-shaped objects, usually pipe. The jaws of the wrench are set in such a way that the teeth bite into the material. The wrench can't be used backward, for it won't grip. Care should be exercised when using the wrench on finished material because it may cut into the material. The wrench is manufactured in various sizes, from 6 in to 48 in. Extra leverage, gained by lengthening the handle, should not be applied. If you need more leverage, use the next size larger wrench. As the wrench becomes older the teeth will round off.

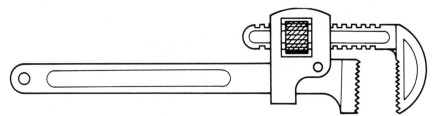

Figure 1-11. Pipe wrench.

This problem can be solved by taking a three-cornered or triangular file and sharpening the teeth back to their normal triangular shape.

Allen Wrench

Socket headed wrenches (Fig. 1-12) or allen wrenches as they're commonly called are used with allen headed screws and bolts. There are two styles of allen wrenches: a short handle and a long handle wrench. They're in all sizes and are always L shaped.

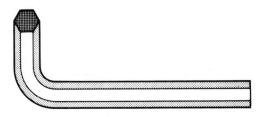

Figure 1-12. Allen wrench.

Special Wrenches

There are many other wrenches for special applications, such as *spanner* wrenches (Fig. 1-13), both fixed and adjustable. There are also end spanner wrenches, pin spanners (Fig. 1-14), and face pin spanners. These are all used to turn special types of nuts with slots cut in the circumference of the nut or with pin holes in the face of the nut.

After looking at the many styles and types of wrenches, you may wonder how you can decide which wrench to use. A few

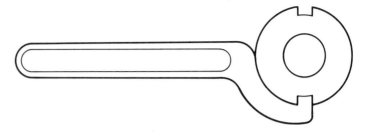

Figure 1-13. Spanner wrench.

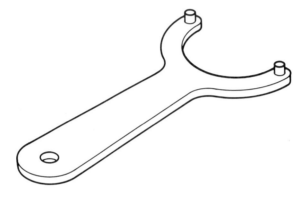

Figure 1-14. Pin spanner wrench.

thoughts ahead of time can be a big help: the job itself, its location, or the number of fasteners to be loosened or tightened. The greatest factor will be experience. After a time on the job the repairman will consider which tool is the best almost instinctively. The old saying "practice makes perfect" holds true in this case.

CHISELS

A chisel is a tool made from tool steel, usually having six or eight sides. One end is blunt and shaped to receive hammer blows. The other end is usually shaped for the operation it's to perform. The chisel is usually formed by forging to the correct shape, heat treating, then forming the cutting edge.

A chisel usually goes through a series of heat treatments to make it strong. Heat treatment helps the cutting edge maintain its sharpness, and at the same time prevents it from breaking or cracking easily. The cutting end is hardened, with the other end being relatively soft to prevent small pieces from spalling or breaking off when being struck by a hammer. Any chisel after moderate use will begin to mushroom in the area that the hammer strikes. This condition can be corrected by grinding off the burrs on the mushroomed part of the chisel and returning it to its correct shape. If this condition isn't repaired, the material will break off when struck by the hammer and small pieces of steel may fly and injure someone.

After moderate use the working or cutting edge of the chisel will become dull and must be resharpened. This is best done on a grinder with a smooth wheel or with a flat file. The chisel can be reshaped and properly beveled. The grinder must be used carefully, for overheating the chisel, evidenced by a bluish color, will draw the temper and render it unsuitable for further use.

The most common chisel is called a *cold* chisel (Fig. 1-15), which is used to cut or shape metals that aren't heated in a furnace. It's used primarily to cut rivets, cut sheet metal, chip small pieces of metal, etc. The size of the chisel is determined by the width of the cutting edge.

There's another chisel called a *hot* chisel but isn't hand held. It's formed like a hammer head, having a handle also. The chisel end is on one end, with a flat end on the other. It's used for cutting or shaping metals that are red hot from being heated in a furnace.

Three other chisels that repairmen should be familiar with are the *cape* (Fig. 1-16), *roundnose* (Fig. 1-17), and *diamond point* (Fig. 1-18) chisels. These three are used for working in grooves and keyways or in corners.

The safety of working with chisels is important to the repairman for he can injure himself or others. The force of the hammer blows can cause small pieces of the chisel or of the workpiece to fly off, possibly striking someone. The importance of keeping the chisels in good shape cannot be overstressed. You should always wear safety glasses or goggles to protect your eyes. Inspect the chisel to insure that there are no mushroomed edges. You should also always hold the chisel correctly, not having too much of an angle between the workplace and the chisel. The thumb and forefinger should always be relaxed, for if the hammer misses the chisel, the injury won't be as painful. When using a chisel on a small workpiece, clamp the workpiece in a vise,

Figure 1-15. Cold chisel.

Figure 1-16. Cape chisel.

Figure 1-17. Roundnose chisel.

Figure 1-18. Diamond chisel.

chiseling toward the stationary jaw of the vise. When using a chisel always work with the chisel directed away from, never toward yourself.

PUNCHES

A group of tools similar to the chisels are the punches. Punches also have an end to be struck by a hammer, and an end to accomplish a type of work.

Center Punch

The most common punch is the center punch (Fig. 1-19). It's most frequently used to punch a small starting hole for a drill bit. This'll start the drill bit in that exact spot. It you don't punch a starting hole first, the drill bit will wander and won't start cutting in the exact place it was intended. The center punch can also be used to make aligning marks on a piece of

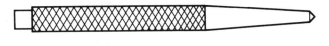

Figure 1-19. Center punch.

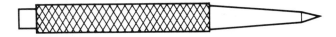

Figure 1-20. Prick punch.

equipment being disassembled. This makes the reassembly easier, for the repairman can line up the marks and properly assemble the equipment. This punch shouldn't be misused, because it's very difficult to repair the point on a center punch. It's possible to repair it on a grinder, but if it's overheated and turns blue, the tip will be too soft to be of further use. When grinding the point, the tip should be left at an approximate taper of 60 degrees.

Prick Punch

A punch related to the center punch is the prick punch (Fig. 1-20). This punch is ground to a very slender point. It's used almost exclusively in layout work to mark the intersection of lines and in some cases the lines themselves. This punch will usually be found only in the shop, for it isn't practical for field work.

Aligning Punch and Drift Punch

The aligning punch and the drift punch are both used for the same purpose, that of aligning holes in material so that a fastener may be inserted. The drift punch is merely stronger than the aligning punch. Some repairmen have a large drift punch with an open end or box end wrench on the opposite end. This is known as a pin wrench, commonly used in heavier work.

Starting Punch and Pin Punch

These two punches (Fig. 1-21) are usually used together. The starting punch is a heavy punch with only a small taper. It's

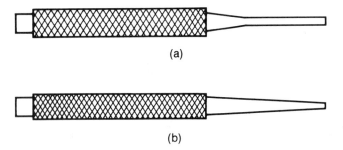

Figure 1-21. (a) Pin punch, (b) starter punch.

used to start driving a rivet or a pin out of a hole. The pin punch has a long, straight, circular shank and is used to finish driving the rivet or pin out of the hole. The pin punch should never be used to start driving the rivet or pin out, because the shank won't withstand the force of the hammer blows. The shank will either bend or break. When using these punches always use the largest diameter possible to prevent damage to the punches.

FILES

A file is made of hardened tool steel, having rows of parallel teeth running diagonally along its surface. The various parts of the file are tang, face, and the tip. All files are divided into two classes, single and double cut (Fig. 1-22). Single cut files have one set of diagonal cutting teeth, while the double cut has two sets of teeth, running in opposite directions. Double cut files remove the material faster, but leave a rough finish. The single cut file is a slower working file but leaves a smoother finish. Both files are available in six different grades. The grades are from dead smooth to rough. The rough file has a large pitch (distance between the teeth) where the smooth file has a very small pitch. The grades are as follows:

1. Rough
2. Coarse
3. Bastard
4. Second cut
5. Smooth
6. Dead smooth

Most files are available in either class and in any grade mentioned. Some of the more common files are listed below. (See Fig. 1-23.)

1. Mill
2. Flat
3. Square
4. Round
5. Half-round
6. Triangular

There are other types of specialty files such as the knife file, pillar file, crossing, etc. These won't, in most cases, be needed by repairmen. The files should be specified by length, type, grade, and class, for example: 12-inch, square, rough, single cut file.

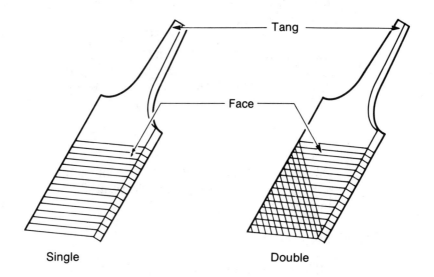

Single   Double

Figure 1-22. Files.

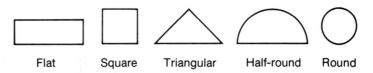

Flat   Square   Triangular   Half-round   Round

Figure 1-23. Some of the more common file shapes.

When using a file, you should always have a handle over the tang of the file. This will prevent any injury, for if the file should slip during use, the tang would be jammed into your hand. When in use, the file should be held in such a manner that it'll give a flat-finished surface. The pressure need not be excessive, but should only be enough to keep the file cutting the material. The push stroke occurs when the pressure should be applied. During the return stroke, no pressure should be applied, for this will dull the teeth on the file. The strokes shouldn't be fast, for this will heat up the file and the material, ruining both.

The file cuts best when the maximum number of teeth are in contact with the material. After continued use the teeth will begin to clog with material. If the cutting action becomes slow, this should be checked. A file card should be used to clean out the teeth. If a file card is unavailable, then a wire brush should be used. The three things that determine the cutting ability of a file are the shape of the file, the sharpness of the teeth, and the hardness of the teeth.

In considering safety of file use, you need to remember what material it's made of: brittle, hardened steel. This means it should never be pryed with, hit with a hammer, or hit against an object. The file would shatter, sending pieces in all directions.

PLIERS

The word pliers implies more than one tool, but it's a plural word used to describe a single tool. There's almost an endless variety of pliers.

Combination Pliers

The most common is the slip joint pliers or combination pliers (Fig. 1-24). These are used where it's unsafe or inconvenient to use your hands. One point to remember is not to use pliers anywhere you could use a wrench.

Channel Locks

The pliers that have grooved holding joints or channels are commonly called "channel locks" (Fig. 1-25). These are very popular with most repairmen, because of the adjustability of

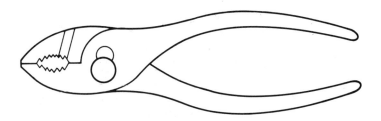

Figure 1-24. Combination pliers.

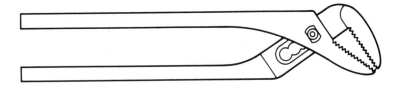

Figure 1-25. Channel lock pliers.

the pliers. They can be used to work on pipe or fasteners of small or medium sizes. These channel locks are not to be used on bolts or nuts when a wrench could be used, because the channel locks will slip or bite into them, making the nuts or bolts unsuitable for re-use. The teeth on the pliers will occasionally become rounded or dull. This can be corrected by working the teeth with a three-cornered or triangular file, till they become sharpened.

Tin Snips

When working with sheet metal, the repairman will use tin snips or aviation snips. Tin snips come in three types: circular, straight, or combination snips. The snips cut the patterns that their names describe, the combination being the most popular, since it'll cut both patterns. The aviation snips are coded, the handle color describing the cut they'll make. The red handles cut to the left, the green cut to the right, and the yellow cut straight patterns.

There are several varieties of wire cutters that fall in the pliers category. These are the diagonal cutting pliers, the lineman's pliers, and the needlenose pliers.

TAPS

A tap (Fig. 1-26) is a cylindrical length of steel with threads around it and flutes running lengthwise along it, forming cutting edges with the threads. A tap is used to cut internal threads. The flutes are provided for the metal chips to fall through, preventing the tap from jamming up when in use. The tap must not be turned all the way into the hole at one time. It should be turned one or two turns, then backed up one-half a turn. This breaks up the metal chips being cut, allowing them to fall through the flutes. This method of tapping will also prevent the tap from jamming and being broken off in the hole.

A complete set of taps is composed of three different taps of the same size. The *taper* tap is used to start cutting the threads. If it's a through hole, it's the only tap required. If the hole is blind or has a bottom to it, you'll need a plug tap which will cut the threads close to the bottom. If you must tap the threads completely to the bottom of the hole, you must use a bottom tap, which is designed for that purpose.

Serial taps are also in sets of three taps. The serial taps are used for tapping in hardened metals. The number of the tap is identified by the number of rings around the top of the tap. The first tap is used to rough-cut the thread. The number 2 tap cuts closer to the proper size. The number 3 tap finishes the thread to proper specifications.

The *extension* tap is merely a regular tap with a long handle on it for cutting threads in deep holes or in areas inaccessible with regular taps.

There are two grades of taps made, cut thread and ground thread. The cut thread is machined to size and then heat-treated. The tap may undergo some changes during heat treatment that may affect the size, so it won't be as precise as the ground thread tap. The ground thread tap is heat-treated then finished to size, which is more expensive but more accurate.

A small tap is turned by a small wrench called a "T" handle wrench. Larger taps are turned by a tap wrench. In the field most repairmen use an adjustable wrench to turn a tap. In using that method, care must be exercised to keep the tap perpendicular to the work. Cutting oil or other suitable lubricants should be used during the tapping operation.

The most skillful repairman occasionally breaks a tap off in

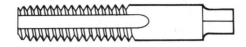

Figure 1-26. Tap.

the workpiece. If the tap extends out of the hole, it's possible to grasp it with pliers and back it out of the hole. However, if the tap is broken off below the surface of the work, then a tap extractor must be used. The extractor is an instrument with long fingers, which extend into the flutes of the tap, to be used to back out the broken part. The fingers are fragile and if too much force is exerted they'll break off in the tap flutes. Great care must then be used to prevent that from occurring. Cutting oil will also make it easier to remove a broken tap.

SAWS

The most common saw that a repairman uses is a hacksaw (Fig. 1-27), designed to cut metal. Most hacksaws are made with an adjustable frame designed to use blades that are 8, 10, or 12 inches long. Hacksaw frames are also adjustable, so that the blade can be held vertically or horizontally. Some manufacturers make the hacksaw to hold the blade at various other angles, allowing the repairman to work in areas that he ordinarily couldn't with a two-position saw.

When installing a hacksaw blade, it should be tightened till the blade is tightly stretched. The teeth should always point away from the handle. This is important, for some saw blades have starting teeth in them, which makes the job easier to start.

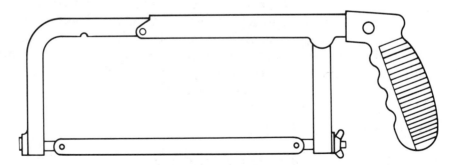

Figure 1-27. Hacksaw.

These teeth are always at the end of the saw away from the handle. Once they've been used to start the cut, then the rest of the blade can be used in long even strokes. You must exert enough pressure to keep the teeth cutting the material. When too little pressure is used it only dulls the teeth on the blade. When the teeth have removed enough material, the slot made by the blade will help to guide it, then pressure should be relieved on the blade during the back stroke. You should move at the rate of fifty strikes per minute; too fast a rate will heat the blade. Overheating will draw the temper and ruin the blade. When a blade breaks during a cut, it is advisable to turn the material over and start a cut from the other side. The width of cut is different from the old to new blade, so turning the material over prevents the blade from binding up. When cutting with a hacksaw, note the sound the cut makes. As you finish cutting through the material, the pitch of the sound changes, warning you to lighten the pressure you are applying during the cutting. This prevents you from cutting through all at once, perhaps losing your balance and striking some part of your body on the work or work table.

STEEL RULE

The steel rule and its variations are given extensive use by repairmen. The rule is used to measure almost everything. The rule comes in various sizes and types, from a nonflexible six-inch steel rule to a fifty-foot flexible tape.

The rule is divided into units of feet and inches. Each inch is subdivided into eights of an inch. The eights are subdivided into sixteenths, thirty-seconds, and finally into sixty-fourths. A steel rule twelve inches long has divisions of eights and sixteenths on one side and divisions of thirty-seconds and sixty-fourths on the other side. If the repairman needs any finer measurement, another instrument should be used. However, the steel rule is sufficient in most applications.

SUMMARY

Hand tools are very essential to our industrialized society. They form the basis for maintaining the equipment of our modern society. Without these basic tools our industry grinds to a stop.

In this chapter we've discussed some of the more frequently used hand tools. Space prohibits the coverage of all tools. The tools discussed will enable the repairman to carry out his job assignments.

Hand tools require certain care if they're to provide maximum service. All hand tools need periodic inspections to determine if they need repair. If repair is required, the repairman should perform it as soon as possible. The better care taken of the tools the longer they'll last.

Hand tools are to be used, not abused. If misused, not only the tool may suffer damage, but also the one abusing it. Safe and proper use will benefit not only the repairman, but also his tools.

REVIEW QUESTIONS

1. Name two types of hammer and the use of each type.

2. Describe the maintenance required on hammers.

3. What are the two main types of screwdriver tips?

4. What is the most important consideration when selecting a screwdriver?

5. What is the proper repair procedure for a screwdriver with a damaged tip?

6. What are three types of fixed-size wrenches?

7. Describe a socket set and its use.

8. Describe an adjustable wrench.

9. What maintenance is required on a cold chisel?

10. List the different types of punches.

11. What's the difference between a single-cut and a double-cut file? What can each be used for?

12. Why should channel locks not be used on a bolt or nut?

13. List the three taps in a complete set.

14. When breaking a hacksaw blade during a cut, why should the part be turned over and a cut made from the other side?

Chapter 2

Power Tools

Hand tools have always been used on equipment repair. To reduce the time involved in some repairs, power tools were developed. Power tools make the job easier and less monotonous for the repairman. The following chapter examines some of the more common power tools. The material presents the popular terminology in dealing with these common tools. Safety is mentioned in an attempt to alert the repairman to the dangers involved in using power tools.

ELECTRIC DRILL

One of the most common tools that a maintenance repairman works with is the electric drill (Fig. 2-1). This tool has many uses; however, some basic safety should be discussed at this point.

When using this tool you should be sure that it's connected to the correct power source. The most common voltage is 110, but you should check the drill before plugging it into an outlet. Don't take for granted that just because the plug fits that it's correct. The person using it before you could have installed an incorrect plug on the drill.

A second point to consider is the condition of the drill housing.

Figure 2-1. Electric drill.

If it's damaged, then you could be subjected to electric shock due to the breakdown in insulation in the damaged housing. One style of drill that's particularly susceptible to this type of damage is the double insulated drill (Fig. 2-2). This style of drill has no ground plug, but has two separate housings to insulate the user from the hazards of the electric current.

Anytime an electric drill is used, you should use safety glasses to prevent injury to your eyes. The drill bits are made of hardened steel and are very brittle. The possibility of the drill bit fracturing and sending projectiles at the operator is very real. You also can reduce this hazard by using only sharpened drill bits.

Parts of the Electric Drill

The basic parts of a standard electric drill are displayed in Fig. 2-3. The *chuck* on the drill is used to hold the drill bit in the

Figure 2-2. Double insulated electric drill.

drill. A chuck key is used to tighten the chuck to prevent slippage. The handle usually has the trigger type switch that controls the drill motor. The switch can be the on and off type or the variable speed type. The variable speed type is controlled by depressing the switch into the handle. The further it's depressed the faster the drill motor runs.

In the diagram (Fig. 2-3), you can see that the drill motor turns a shaft connected to a small set of gears that can reduce the speed of the output, but at the same time increase the output torque.

In selecting the proper drill bit, it's important to realize that the bigger the drill bit, the slower the drill should turn. Conversely, the smaller the drill bit the faster the drill should run. This is why the variable speed drills have a definite advantage in field use. Some of the recommended drill speeds for given sizes are found in Table 2-1.

Drill Attachments

There are many special attachments for electric drills that can help the repairman increase the speed and ease with which he performs his job. One such attachment is the *holesaw*. The holesaw (Fig. 2-4) uses a small drill bit to start the hole and a larger thin-walled piece of steel to cut the larger hole. The hole

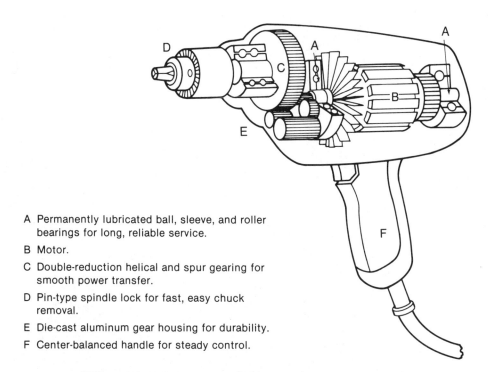

A Permanently lubricated ball, sleeve, and roller bearings for long, reliable service.

B Motor.

C Double-reduction helical and spur gearing for smooth power transfer.

D Pin-type spindle lock for fast, easy chuck removal.

E Die-cast aluminum gear housing for durability.

F Center-balanced handle for steady control.

Figure 2-3. Basic parts of the standard electric drill.

Figure 2-4. Hole saw attachment.

Table 2-1   Drill Bit Speeds

Drill Bit Size in Inches	Drill Bit Speed (in RPM)
$\frac{1}{16}$	2000 RPM
$\frac{1}{8}$	1200 RPM
$\frac{3}{16}$	820 RPM
$\frac{1}{4}$	600 RPM
$\frac{5}{16}$	490 RPM
$\frac{3}{8}$	400 RPM
$\frac{7}{16}$	350 RPM
$\frac{1}{2}$	300 RPM
$\frac{5}{8}$	250 RPM
$\frac{3}{4}$	200 RPM
1	153 RPM

saw is available in many sizes and is very useful in working with thin metals and woods. Care must be exercised in using the hole saw so as not to exert excessive force and actually break the outer shell.

The *countersink* (Fig. 2-5) can be used to drill holes for bolts that must have their heads recessed below the surface of the material in which they're inserted. These'll usually be flat-head machine screws or allen screws. This is particularly good for two surfaces that must be assembled flush or must slide on one another.

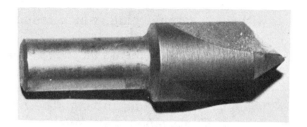

Figure 2-5. Counter sink.

POWER SCREWDRIVERS

Power screwdrivers (Fig. 2-6) or *nut drivers* (Fig. 2-7) are also useful in applications where large quantities of nuts or screws must be tightened. Again in this application a variable speed drill is particularly useful.

Figure 2-6. Power screwdriver bits.

Figure 2-7. Power nut drivers.

ELECTRIC HANDSAWS

Electric handsaws are another group of tools with which a repairman needs to be familiar. There are many types and sizes of handsaws. Figure 2-8 illustrates a basic design of a power saw. The saws are also available in double-insulated models without a ground plug, which reduces the hazard of electric shock.

There are adjustments on the saw that allow for the depth of cut and also the angle of cut (A), for making beveled edges.

A variety of blades may be used on the saw. The *crosscut blade* is used for fine cutting. The *rip blade* is used for rough cutting. The *combination blade* applies the best characteristics of both; it uses less power to run and is easier to sharpen.

While the power saw is very helpful to the repairman, it can also be very dangerous if used improperly. When in use you should keep the saw away from your body, and you should try to keep both hands on the saw. This'll keep you from getting your hand in the blade path. As unbelievable as it may seem, there have been incidents where unthinking individuals have put their hands under the work to feel if the saw blade is coming through. It's difficult to reattach fingers removed with a power saw, due to the rough cut. With a power saw, you shouldn't try to cut a curved line, but instead should try to keep all cuts straight. If the blade is in a bind, it can fracture or shatter.

SABER SAW

If you're going to try to cut odd shaped figures, you should use a saber saw (Fig. 2-9). The saber saw can be used in cutting all sorts of irregular shapes. When using it you shouldn't force

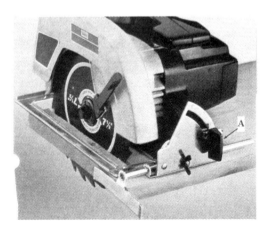

Figure 2-8. Electric hand saw.

Figure 2-9. Saber saw.

it into the work, but should let it cut and feed it into the work slowly.

You can use a variety of different blades with the saber saw. They all fall basically into three categories: wood cutting, metal cutting, and knife.

The wood cutting is used for wood and usually has six to twelve teeth per inch. The metal cutting blade has six to thirty-two teeth per inch, depending on the thickness of the metal. The knife blade is used for cutting paper, leather, or cloth. It's usually a smooth blade without teeth.

The saber saw should be treated like the power saw, and given the utmost care when in use. Any power tool has the potential to maim or kill when in use, so safety should be a foremost consideration.

GRINDERS

The next tool that we'll consider is the grinder. The grinder can be powered electrically or pneumatically (Fig. 2-10). No matter how the grinder is powered, safety is very important. The grinder should always be connected to the proper power supply. Care should be taken to insure that the air grinder is

Figure 2-10. Grinders.

connected to air. In some industries the air, water, and oxygen lines all look the same. To connect the grinder to an improper source could not only damage the tool but also result in injury to you or your co-workers.

It's essential to keep the power cord in good repair. A damaged cord results in improper operation of the tool and presents a shock hazard to the workers. The grinder power cord should always be inspected carefully prior to being used. One point that's neglected in inspecting grinders is the *grinding wheel*. Wheels are susceptible to damage by dropping or hitting into the work with excessive force. A small crack or fault in the wheel can cause it to explode into shrapnel. There have been many incidents in industry where workers have been killed or maimed by exploding grinding wheels. The way to assure yourself of a safe wheel is to ring-test the wheel before it's mounted on the grinder. This is performed by supporting the wheel through the center with a nonmetalic rod or shaft. Then with a wood handle of a hammer, or some other suitable instrument, you should lightly strike the wheel at two positions about 45 degrees either side of top dead center (Fig. 2-11). If the wheel has a high pitched ring, it's safe to use. If it has a dead sound or a metallic thud, it shouldn't be used because it has a crack or a defect in the wheel. This test should be run each time a wheel is mounted, for new wheels can be damaged in transit. Even after the test is run, and the wheel mounted, the grinder should be held away from the body when started, so that if something happens you would be in the clear.

One last consideration in grinder safety is the proper use of guards. Many times they're left off because it makes it more

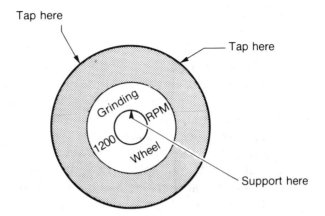

Figure 2-11. Ring testing a grinding wheel.

difficult to work with the grinder. This is a poor practice for this exposes the workman and his co-workers to unnecessary dangers. Guards are on there for a purpose—keep them on the grinder. Always direct the grinding spray away from your body, and away from others in the area.

Three styles of grinder—vertical, horizontal, and pencil—are shown in (Fig. 2-12).

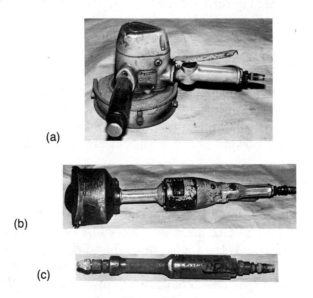

Figure 2-12. Vertical, horizontal, and pencil grinders.

IMPACT WRENCH

No repairman's tool list is complete without an impact wrench. A standard impact wrench is pictured in Figure 2-13. The impact wrench is air driven through a small reduction. It's usually equipped with a directional switch so that it becomes reversible. It comes in several sizes from very small to some drives that may be an inch or larger. This tool is particularly useful when parts may have to be assembled and disassembled in large quantities. One point to keep in mind is that you should use only impact sockets with the impact wrench. Standard sockets will eventually fracture if used on an impact wrench. One maintenance point to keep in mind is that occasionally you should add a small amount of oil to the air supply to keep the air drive mechanism free and in good condition.

PNEUMATIC HAMMER

The pneumatic hammer (Fig. 2-14) is used on occasion to remove rivets or perform large quantities of chiseling. One safety tip that applies to this tool that doesn't apply to others is that you must never point it at anyone. The chisel tip or any other tip for that matter can fly out of the spring holder and can injure bystanders.

PNEUMATIC DRILL

In some special cases you may use a pneumatic drill. One such drill is pictured in Fig. 2-15. All safety rules that apply to electric drills and pneumatic tools apply to this tool also. As with any power tool care should always be exercised in its use.

SUMMARY

In summary, power tools can be of great benefit to any repairman. The tools must always be connected to the proper power source. If the tools are mishandled, a very real danger is present. If care is taken with power tools, they can improve the productivity of the repairman. This again enables the repairman to work smarter, not harder.

Figure 2-13. Impact wrench.

Figure 2-14. Air (pneumatic) hammer.

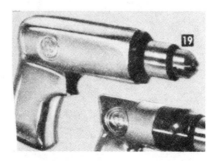

Figure 2-15. Pneumatic drill.

REVIEW QUESTIONS

1. To what danger can damage to the housing of a double insulated drill expose the repairman?

2. The part of the drill that holds the drill bit is called the _____.

3. What advantage does the variable speed drill have over the standard drill?

4. What is a holesaw used for?

5. To install a bolt head flush with a machine part, what attachment should you use?

6. What are the three types of blades found on a power saw, and what are each used for?

7. A power saw should be used to make only _____ cuts.

8. A _____ saw should be used to make irregular cuts.

9. Before any grinding wheel is used, it should be checked by a _____ test.

10. Why should standard sockets not be used with an impact wrench?

Chapter 3

Fasteners

A study of items important to maintenance repairmen wouldn't be complete without looking at screw threads and fastening devices.

SCREW THREADS

Screw threads are used in industry to fasten parts, hold pieces in position, and to transmit power. The threads may be *internal*, such as a tapped hole or in a nut; or they may be *external* such as a bolt or a stud. No matter which you may have, they all have terms in common. Following is a list of terms relating to threads.

1. **Root** of the thread is the base of the thread between two adjacent threads (Fig. 3-1).

2. **Major diameter** is the term applied to the largest diameter of the screw or nut.

3. **Minor diameter** is the smallest diameter of the bolt or nut. This term may be used interchangeably with root diameter (Fig. 3-1).

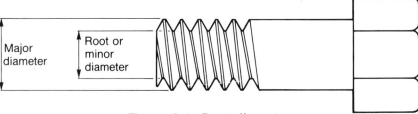

Figure 3-1. Root diameter.

4. **Pitch diameter** is the major diameter minus the minor diameter, plus the minor diameter (Fig. 3-2).

5. **Pitch** is the distance from one point on a thread to the same point on the next thread, measured axially along the threads.

6. **Lead** is the distance the screw thread moves axially in one full turn. On a single thread, the lead is equal to the pitch. In multiple threads, the lead is equal to the number of thread starts times the pitch. (For example, double thread equals 2 times the pitch.) (Fig. 3-3)

7. **Multiple threads** is the term used to describe a thread that has more than one starting thread. It's actually like having two or more sets of threads on a fastener. The advantage gained is that you can move a mating nut farther in one turn on a multiple thread than on a single thread without sacrificing the strength of the unit. It'll travel the number of thread starts (2, 3, 4, etc.) times the pitch. These are usually used on units needing faster travel along the threads' axis than can be obtained on a single thread. (See Fig. 3-3)

8. **Number of threads** refers to the number of threads along a 1-inch distance along the threads' axis (Fig. 3-4).

9. **Length of engagement** is the length of contact between two mating threaded parts, measured axially.

10. **Depth of engagement** is the depth of contact between two threaded mating parts, measured radially.

11. **Fit** is the term used to describe the amount of clearance in mating threads. It has a standard designation as follows:

1 = loose fit   A = external threads
2 = medium fit   B = internal threads
3 = tight fit

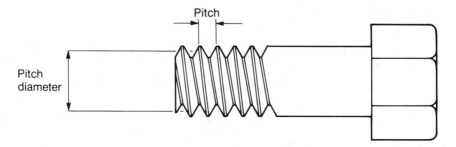

Figure 3-2. Pitch.

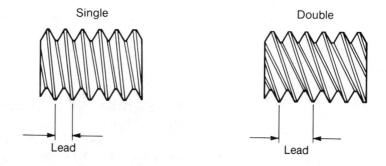

Figure 3-3. Leads of single and double threads.

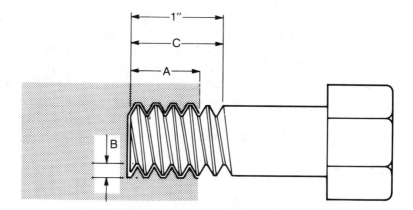

A—Length of engagement.
B—Depth of engagement.
C—Number of threads per inch.

Figure 3-4. Number of threads.

FASTENERS

The most common use of screw threads is in fastening devices such a bolts, studs, and screws.

This is a good place to discuss the correct terminology applied to threaded fasteners. A *bolt* is used with a *nut* for tightening. If the fastener is threaded into a hole and tightened by turning the head of the fastener, it's called a *screw*. If it's threaded on both ends, it's called a *stud*. (See Fig. 3-5.) The majority of these fasteners use either the Unified National Coarse (UNC) or the Unified National Fine (UNF) thread designations. The vast majority of the threaded fasteners use the UNC designation. The UNF designation is used mainly in automotive and aeronautical work.

BOLT GRADES

There are many materials used to make threaded fasteners, and a detailed description of the metallurgy involved would fill many volumes itself. Let us emphasize the differences by looking at a table of bolt grades (Table 3-1). We can see the various strengths of the bolts, but keep in mind that the stronger the bolt the more expensive it is. Use only the strongest grade you need for the job; not all applications require a grade 8 bolt. If they did, that would be all that's manufactured. The most popular grades of threaded fasteners are the low carbon (grade 2), carbon heat treated (grade 5), and alloy heat treated (grade 8). There's a code for marking the fasteners. The head of the fastener has raised radial lines. The code for the grade of the bolt is the number of lines plus 2. So a grade 5 would have three radial lines on its head. The more radial lines the greater the strength of the bolt.

NUTS

Nuts are internally threaded fasteners used with bolts, studs, or other externally threaded fasteners. Nuts come in three common grades: grade 2, grade 5, and grade 8. The *hex nut* (Fig. 3-6) is the most common nut used in fastening applications. The *hex jam nut* is also used in the holding action of fasteners. The hex jam nut is run down against the workpiece and the stan-

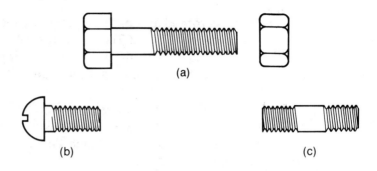

Figure 3-5. (a) Bolt, (b) screw, (c) stud.

Table 3–1

Grade Marking	SAE Number	Tensile Strength
	2	64,000 psi
	5	105,000 psi
	6	130,000 psi
	8	150,000 psi

dard hex nut is forced against it, effectively locking the threads.

The *hex castle nut* (Fig. 3-7) is used in applications where locking the fastener is important. The nut is equipped with a slot for insertion of a pin to lock the fastener. The bolt or stud must also be drilled for the cotter pin.

There are also *square nuts* (Fig. 3-8). They come in standard and heavy series. The heavy series are wider across the flats and also thicker than the standard series. These nuts are usually used with square-headed bolts.

Wing nuts (Fig. 3-9) are used when a part is to be assembled and disassembled frequently. The assembly can use only the pressure generated by the fingers. For any assembly requiring more holding force you should use a standard nut.

The last group of nuts is classified as *lock nuts*. Lock nuts are divided into two categories: prevailing torque and free-spinning. The *prevailing torque* requires torque to run it down the threads. A common type has a synthetic insert that'll resist loosening after the proper torque has been applied. A *free spinning lock nut* runs on easily but deforms under torque and thus the threads are locked.

Figure 3-6. Hex nut.

Figure 3-7. Hex castle nut.

Figure 3-8. Square nut.

Figure 3-9. Wing nut.

WASHERS

Washers are also important in fastening applications. They're divided into two classifications: flat washers and lock washers (Fig. 3-10).

Flat washers are also called bearing washers for they're used to prevent the fastener from cutting into the material it's holding during tightening. They work especially well on soft materials. The styles are too varied to mention but each manufacturer carries many different styles, so consult your local distributor for any unusual applications.

Lock washers are hardened pieces of steel, usually high carbon or alloy steel. The slight deformation in the washer puts extra stress on the nut to prevent loosening. Unfortunately, lockwashers are over-rated and misused. If the fastner is properly tightened it doesn't require a lockwasher to hold it. The torque applied will deform the threads sufficiently enough to lock the fastener. The problem arises when not enough torque is applied to the fastener to deform the threads; then lockwashers are used. If the repairman properly torques the fastner, it'll never require a lock washer.

(a)   (b)

Figure 3-10. (a) Flat washer, (b) lock washer.

ALLEN CAP SCREWS

Before leaving the area of fasteners, we must consider the allen cap screws (Fig. 3-11). These fasteners are very hard, being rated the same as the grade 8 hex-headed bolt. The fastener must be tightened by a hex wrench. They're used primarily in tool and die work, but are becoming more popular where a strong fastener is required. Allen socket screws come in several different styles.

The *flat headed socket cap screw* (Fig. 3-12) is used when flush

Figure 3-11. (a) Allen cap screw, (b) Allen wrench.

Figure 3-12. Flat headed Allen cap screw.

Figure 3-13. Button head socket cap screw.

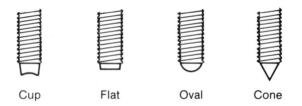

| Cup | Flat | Oval | Cone |

Figure 3-14. Set screw tips.

mounting of an object is important. It may also be used to fasten an object when another object is going to slide on top of it. The head of the bolt being countersunk will prevent any interference.

The *button head socket cap screw* (Fig. 3-13) is used where a larger bearing area is required for the fastener.

Included in the classification of allen cap screws is the allen set screw. They're usually threaded into a tapped hole for the purpose of fastening or securing two parts together or to prevent movement of mating parts. They come in many different point styles such as cup (most popular) flat, oval, and cone (Fig. 3-14).

One last point to consider in fasteners is the plastic locking materials such as Loctite® (manufactured by the Loctite Corporation). This is not a commercial endorsement of the product, but it does have its place in mechanical fasteners. When applying the proper grade of locking material correctly, unbelievable holding force can be generated. The important point to remember is to use the lightest grade that'll work, and work up to the thicker grades. The good thing is that Loctite is impervious to solvent. The only thing that can break a properly applied bond is heat. It cannot be reused. Once the bond is broken new material must be applied.

SUMMARY

In summary, the field of fasteners is an ever growing field. Special requirements will produce a new group of fastening devices as industry continues modernization. It's up to the repairman to keep abreast of new inventions in this field.

REVIEW QUESTIONS

1. Define the difference between pitch and lead.

2. What is the difference between a bolt and a screw?

3. How can you distinguish between a grade 2 bolt and a grade 8 bolt?

4. What advantage does a castle nut have over a standard nut?

5. What types of applications are wing nuts used for?

6. How does a lock washer prevent a fastener from loosening?

7. What type of locking material is a good alternative to a lock washer?

Chapter 4

Basic Mechanics

MASS

One of the first terms we need to consider is mass. Mass can be simply stated as the measure of the amount of material in a body independent of the gravitational attraction of the earth. The accepted unit of measure is the pound; however, when investigating further into more detailed dynamic mechanics, we find that the unit is actually lb-sec²/ft. The formula for finding mass is:

$$m = \frac{f}{a}$$

m = mass
f = force (weight)
a = acceleration due to gravity (32 feet/second squared)

A sample problem might read like this: calculate the mass of a 20,000 pound coil of steel.

mass = ?????
force = 20,000 lbs
accel = 32 ft/sec²

$$\text{mass} = \frac{20,000 \text{ lbs}}{32 \text{ ft/sec}^2}$$

Inverting and multiplying:

$$\text{mass} = \frac{20,000 \text{ lbs·sec}^2}{32 \text{ ft}}$$

$$\text{mass} = 625\frac{\text{lbs·sec}^2}{\text{ft}}$$

WEIGHT

Weight is the amount of gravitational attraction the earth has for an object. Weight is also measured in pounds. It can be distinguished from mass by the following illustration. If an object weighs 100 pounds on earth and is transported to outer space, it's weightless. The amount of material in the object (mass) has not changed; only the gravitational attraction exerted by the earth has changed. So the difference between mass and weight becomes apparent.

Weight can be considered to be a force. The earth's surface acts upward with an equal force to keep the object from being drawn into the center of the earth. One other force that should be considered is friction.

FRICTION

Friction is the resistance encountered when you attempt to roll or slide one object on another. For motion to occur, the applied force must be larger than the opposing frictional force. There's a definite relationship between the types of material and the frictional resistance. This relationship is represented by the coefficient of friction.

Coefficient of Friction

There are two types of coefficient of friction, static and dynamic. The *static coefficient* is the object's resistance to sliding divided by the force pressing the two objects together (usually the weight). Looking at the formula, we see:

Cf = Rs/Fp.
Cf = coefficient of friction
Rs = resistance to sliding
Fp = the force pressing the two objects together.

The *dynamic coefficient* relates the sliding frictional force to the weight of the object. Since it takes less effort to keep an object moving once motion has begun, the dynamic coefficient will be less than the static coefficient. Looking at the formula:

Cd = Fs/Fp
Cd = dynamic coefficient
Fs = sliding frictional force
Fp = the force pressing the two objects together.

The relationship between the two coefficients can be seen in Table 4-1.

Table 4-1   Coefficient of Static Friction
(of Clean, Dry Materials)

Materials	Coefficient
Steel on steel	0.8
Steel on brass	0.35
Steel on graphite	0.1
Steel on teflon	0.04
Aluminum on aluminum	1.35
Glass on metal	0.6
Iron on iron	1.0
Copper on copper	1.0
Wood on wood	0.25–0.5
Leather on metal	0.6

WORK

If the applied force can overcome the resistance, then motion will take place. When the force is moving through a distance, then work is performed. If the force is applied and no motion occurs, then no work is accomplished. This fact can be illustrated by the following formulas:

W = F × D
W = work
F = force
D = distance

If the force is applied and no motion occurs than you have:

W = 100 lbs. × 0
W = 100 × 0
W = 0

So, for work to be performed motion must occur. Work can be expressed in any units of force and distance. If we measure the time during which work is performed, then we can measure the power used:

P = F × D/T

Work is usually measured in ft-lbs/min or ft-lbs/sec.

Horsepower

In industry, power is consumed in such large quantities that it's difficult to work with it in these small units. Horsepower is used to measure large quantities of power. A horsepower is equal to 33,000 ft-lbs/min, or 550 ft-lbs/sec. The formula is:

Hp = F × D/T × 33,000
 F = force in lbs.
 D = distance in ft.
 T = time in min.

When we look at transmitted power, we must consider the efficiency of the transmitting medium.

$$\text{Eff.} = \frac{\text{Power output}}{\text{Power input}} \times 100\ \%$$

Efficiency

Efficiency is the ratio of the power output divided by the power input times 100%. No device is 100% efficient, so the efficiency ratings are always less than 100%.

Inertia

When we discuss motion, we must consider Newton's first law of motion, which is also called the "law of inertia." Inertia is the property of a body that requires the application of an outside force to accelerate or decelerate the body. The formula is:

a = Vf − Vo/T
a = acceleration
Vf = final velocity
Vo = original velocity
T = time
d = Vo − Vf/T
d = deceleration

Note: velocities must be in feet per second, time must be in seconds, and the acceleration or deceleration is in feet per second squared (for the above formulas).

Velocity

Other terms are as stated above. Velocity is merely the distance divided by the time it takes to cover the distance. The formula for velocity is:

$$\text{Velocity} = \frac{\text{distance}}{\text{time}}$$

This acceleration and motion that we've been discussing so far is linear or straight line motion. In dealing with power transmission, we also need to consider rotary velocity and motion.

Rotary Velocity

Rotary velocity is usually measured in rpm (revolutions per minute). There's a definite relationship between rotary velocity and linear velocity. The conversion formula is as follows:

V1 = .262 × D × RPM
V1 = Linear velocity in feet per minute
D = diameter in inches

RPM = revolutions per minutes

Torque

The force that tends to produce rotation is torque. Torque is a force times the distance at which it's acting. Notice that no motion is required to apply torque. Looking at the formula:

$T = F \times D$
T = torque
F = Applied force.
D = the distance at which the force is applied

This difference helps to draw the line between work and torque.

AREAS AND VOLUMES

Two other types of measure that a repairman needs to be familar with are area and volume. It's important to be able to figure the weight of an object by examining its shape. The following is a list of formulas that apply to areas:

Square area = s^2
Rectangle = $H \times L$
Triangle = $H \times B/2$
Circle = πr^2

Volumes of solids are also important in figuring weights. The following is a list of volumes that are frequently used:

Rectangular solid = $L \times W \times H$
Cylindrical solid = Area of end $\times$ Length
Prism = $A \times$ Width

If you know the total volume of a solid in either cubic inches or cubic feet, then you can estimate the weight of the object. The formula is:

$W = V \times D$
W = weight
V = volume
D = density

In using the formula, you need to be sure your units of measure are the same. For example, if you have a block of steel that is an 8-inch square, and want to know the weight, you can use the above formula. The problem would be worked like this:

W = V × D
weight = ????
Volume = 8 × 8 × 8 = 512 cu.in.
density = .283 lb/cu.in.

$$\text{weight} = \frac{512 \text{ cu.in.} \times .283 \text{ lb}}{\text{cu.in.}}$$

weight = 145 lbs

If the object is irregular in shape, then you may have to break it down mentally into several regular shapes (see Fig. 4-1).

Figure 4-1. Example of weight calculations.

EXAMPLE. Estimate the weight of a cast iron roll (solid). Find the weight of the cast iron roll shown.

Roll Dia. = 24″—2′
Roll Face = 42′—3½′
Shaft Dia. = 9″—.75′
Shaft length = 12″—1′

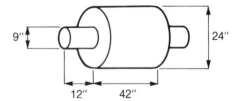

STEPS

1. Find the volume of the roll shafts in cu. ft.
2. Find the volume of the roll body in cu. ft.
3. Multiply the total volume by the density.
   (1a) Volume of shafts
   Volume = End area × length
   Volume = Radius × radius × π × length
   Volume = 3/8 × 3/8 × 22/7 × 1 = 99/224 or .45 cu. ft.
          = 0.375 × 0.375 × 3.1416 × 1 = .4417864669 cu. ft.
          = .45
   Volume of 2 shafts = .45 × 2 = 0.9 cu. ft.

(2a) Volume of roll body
Volume = Radius × radius × π × length
Volume = 1 × 1 × 3.1416 × 3.5 = 11 cu. ft.
(3a) Weight of roll—(cast iron 500 lb./cu. ft.)
Weight = Volume × weight per cu. ft.
Weight = 11.9 × 500 = 5950 lbs.

STEPS (chambered roll) (shafts = 4″ dia. × 15″ long)
(body = 12″ dia.—36″)

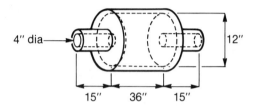

1. Find the volume of the solid shafts. (from previous problem)
2. Find the volume of the solid roll body. (from previous problem)
3. Find the volume of the chambers in shafts. (1728 cu. in. per cu. ft.)

$$0.7854 \; D^2 \times length \times 2 = .2 \; cu. \; ft. \; (0.218166)$$

4. Volume of body chambers

$$0.7854 \; D^2 \times length = .07854 \times 12^2 \times 36 = 4071.5 \; cu. \; in.$$
$$= 4071.5/1728 = 2.3562 \; cu. \; ft.$$

5. Subtract the volume of the chambers from the solid volume to get the metal volume.

Solid volume = 11.9   Chambers = .2 + 2.3 = 2.5 cu. ft.
Metal volume = 11.9 minus 2.5 = 9.4 cu. ft.

6. Multiply the metal volume by the weight per cubic ft.

Weight = Volume × Density   Weight = 9.4 × 500 = 4700 lb.

Figure 4-1. Continued

DRIVE RATIOS

One final point to consider is drive ratios. Drive ratios may be divided into three basic types, belt, chain, and gear.

Belt Drive Ratios

Belt drives are based on the size of the pulleys. (See Figure 4-2) The formula for the pulley drive ratio is as follows:

$$O.S. = \frac{I.P.D. \times I.S.}{O.P.D.}$$

O.S. = Output speed (r.p.m.)
I.S. = Input Speed (r.p.m.)
O.P.D. = Output pulley diameter
I.P.D. = Input pulley diameter

For example, if you have a pulley 10 inches in diameter, running at 500 rpm, driving a pulley that has a diameter of 5 inches, what's the output speed?

O.S. = ?????
I.S. = 500 r.p.m.
O.P.D. = 5 inches
I.P.D. = 10 inches

$$O.S. = \frac{I.P.D. \times I.S.}{O.P.D.}$$

$$O.S. = \frac{10 \text{ inch} \times 500 \text{ r.p.m.}}{5 \text{ inch}}$$

$$O.S. = \frac{5000}{5}$$

$$O.S. = 1000 \text{ r.p.m.}$$

Chain Drive Ratios

Chain drive ratios are based on the number of teeth in each sprocket. The following formula illustrates this point:

$$O.S. = \frac{T.I.S. \times I.S.}{T.O.S}$$

O.S. = output speed (rpm)
I.S. = input speed (rpm)
T.I.S. = teeth on input sprocket
T.O.S. = teeth on output sprocket

For example, if a sprocket having 50 teeth running at 400 rpm is driving a sprocket having 200 teeth, what's the output speed in rpm?

O.S. = ???????
I.S. = 400 rpm
T.I.S. = 50
T.O.S. = 200

$$\text{O.S.} = \frac{\text{T.I.S.} \times \text{I.S.}}{\text{T.O.S.}}$$

$$\text{O.S.} = \frac{50 \times 400}{200}$$

$$\text{O.S.} = \frac{20000}{200}$$

$$\text{O.S.} = 100 \text{ rpm}$$

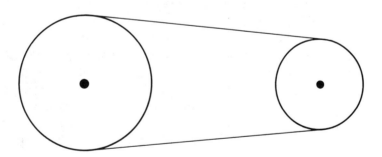

Figure 4-2. Pulley drive.

Gear Ratios

Gear ratios are a lot like the chain ratios except that they use the number of teeth in a gear instead of the number of teeth on the sprocket. The formula is:

$$O.S. = \frac{N.T.I.G. \times I.S.}{N.T.O.G.}$$

N.T.I.G. = number of teeth on input gear
N.T.O.G. = number of teeth on output gear

For example, if the input gear having 20 teeth is turning at 300 rpm, what will the speed of the output gear be if it has 180 teeth?

$$O.S. = \frac{N.T.I.G. \times I.S.}{N.T.O.G.}$$

$$O.S. = \frac{20 \times 300}{180}$$

$$O.S. = \frac{600}{180}$$

$$O.S. = 3.33 \text{ rpm}$$

SUMMARY

Basic mechanics theory seems immaterial to repair equipment. As experience is gained, the relationship becomes more obvious. The basics may be referred to throughout the book, so it's good to understand these points before continuing through the book. In this unit we've discussed the basic theory relating to mechanical drives. We've explored the difference between mass and weight. We've determined that weight is important in determining the friction that results in sliding or rolling two objects together. Work, power, and efficiency are all important in power transmission. If we can't determine efficiency, then we may be using equipment that wastes energy.

REVIEW QUESTIONS

1. Explain the difference between mass and weight.

2. Describe the difference between the static and dynamic coefficient of friction.

3. What is the relationship between work and power?

4. Why is efficiency important to know?

5. Explain the difference between rotary and linear velocity.

6. In what way is torque different from work?

7. How can you determine an object's weight if you know the type of material and its size?

8. If a pulley measuring 10 inches in diameter drives another pulley 20 inches in diameter, what's the speed of the first pulley, if the driven pulley is turning 250 rpm?

9. Why is knowing the drive ratio important?

Chapter 5

Lubrication

PURPOSES OF LUBRICATION

Lubrication is fundamental to proper equipment care. Lubricants are used for the following purposes:

1. To reduce friction.
2. To reduce metal to metal contact, which reduces wear.
3. To provide a metal-separating wedge of lubricant, which dampens shock loads.
4. To dissipate heat.
5. To prevent rust and corrosion.
6. To provide a barrier against contamination.

WAYS LUBRICANTS ARE APPLIED

Lubricants are applied in the following manners:

1. **Gravity or drip** (Fig. 5-1). The lubricant is fed into the lube system by gravitational force, and usually dispensed in small amounts at slower operating speeds.

2. **Splash method** (Fig. 5-2). A slinger or some other device splashes the oil onto the part to be lubricated.

3. **Bath method** (Fig. 5-3). The device is partially immersed in oil and the lubricant is carried throughout the system.

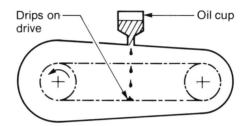

Figure 5-1. Gravity or drip method of applying lubricant.

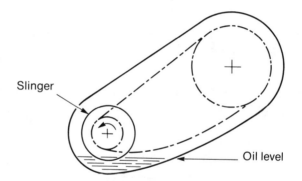

Figure 5-2. Splash method of applying lubricant. Oil is picked out and splashed on drive. (Courtesy of P. T. Components, Inc.)

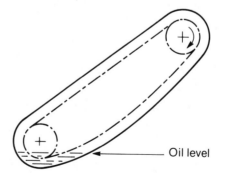

Figure 5-3. Bath method of applying lubricant. Drive is partially submerged in oil. (Courtesy of P. T. Components, Inc.)

4. **Pressure method** (Fig. 5-4). This method is used to spray lubricant in critical areas needing lubrication. Usually a pump dispenses the lubricant.

5. **The manual method.** The lubricant is periodically applied by a brush or some other hand application method.

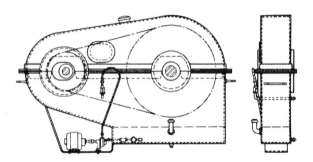

Figure 5-4. Spray method of applying lubricant. (Courtesy of P. T. Components, Inc.)

Lubricants come in divisions based on type. The divisions are as follows:

1 . Liquid—oils of all types
2. Semisolid—all types of greases
3. Solid—metals of all types
4. Gases—used to separate or cool moving surfaces

Understanding lubrication involves understanding the manufacturer's description of the product. The following is a list of common terms:

1. **Additive.** Any substance added to a lubricant to enhance some particular characteristic. Table 5-1 lists some common additives.

2. **Bleeding.** Separation of oil from its base in greases.

3. **Penetration.** A test for greases. A cone is dropped from a given height into the grease. The depth the cone penetrates is the penetration or consistency rating of the grease.

4. **Dropping point.** The temperature at which a grease becomes a liquid.

5. **Flash point.** The temperature that a vapor collected from a fluid will ignite.

6. **Neutralization number.** Measure of the acidity of a lubricant.

7. **Oxidation.** Breakdown of lubricant due to heat and air.

8. **Pour point.** The lowest temperature that a fluid will flow under its own weight. *Weight of oil*

9. **Pumpability.** The measure of a grease's ability to be pumped.

10. **Viscosity.** A fluid's resistance to flow.

OILS

Oils are the most common kind of lubricant. Oils are classified into three categories: mineral oil, animal and vegetable oils, and synthetic oils.

Mineral Oils

Mineral oils are oils drawn from the ground in the form of crude petroleum. The oils are refined and then the necessary additives are mixed in to provide the qualities that the manufacturer is trying to achieve. These additives will be considered in later material.

Animal and Vegetable Oils

Animal and vegetable oils are derived from natural sources. Animal oils usually come from animal fatty tissues and also fish oils.

Synthetic Lubricants

Synthetic lubricants are made of some chemical compound. This then will be used in special application where other oils cannot be used. The problem with synthetic lubricants is that they're very expensive to manufacture. They become useful at high temperatures. Any temperature over 200 degrees begins to break down animal or petroleum based oils. Synthetic lubricants then become a necessity, even at a high cost.

GREASES

Greases are usually classified by the type of soap base used in forming the grease. The common types are:

- Lithium
- Sodium
- Calcium
- Aluminum

Each of the greases has its own qualities, making the choice of the correct grease a challenge. One caution in selecting greases is do not select a too heavy grease. A grease that's too heavy will channel. This means that it won't flow back into the area that needs lubricating. Also a too thin oil won't keep the parts to be lubricated separated, and will allow contact. This usually results in rapid wear and destruction.

SOLID LUBRICANTS

Solid lubricants are metals that are soft and have a very low coefficient of friction when in contact with the materials to be lubricated.

GASES

Gases are used as lubricants in areas where the materials are constructed so as to allow a cushion of air to separate the moving parts. Some bearings using this principle are called *gastatic* bearings. The caution here is that all gases used as lubricants must be extremely clean.

ADDITIVES

Many oils and greases couldn't be used in certain applications if the manufacturer didn't blend in some type of additive. Additives are used to give the lubricants certain qualities that are useful in lubrication. Some of the more common additives are listed in table 5-1.

Table 5-1   Lubricant Additives

Additive	Purpose
Oxidation inhibitor	To prevent corrosion and the formation of varnish and sludge.
Detergent	To prevent the formation of deposits.
Dispersant	To keep deposits in suspension to prevent them from forming on any metal parts.
Extreme pressure	To reduce wear by increasing film strength of the fluid.
Foam inhibitor	To prevent the formation of foam.
Pour point depressant	To allow the oil to pour at lower temperatures.
Viscosity index im prover	To improve the visocosity of the lubricant to prevent breakdown at increased temperatures.
Rust preventative	To prevent the formation of rust during equipment shutdowns.
Water repellent	Usually found in greases, to prevent penetration of water into areas needing lubrication.

Handwritten annotations: cleaners (under Detergent); gummy deposits formed on walls (under Dispersant); increases load carrying capacity of oil (under Extreme pressure); when oil exposed to air (after foam); weight / 10w-40 (under Pour point depressant); "STP" viscosity stays the same as temp increases (under Viscosity index improver)

TOOLS FOR APPLYING LUBRICANTS

Greases are usually applied by a hand gun, power gun, or by an automatic lubricating system.

The *hand gun* (Fig. 5-5) is usually filled from a large container of lubricant. It may be a bucket or barrel of grease. If it's filled from either of these, care must be taken that the grease is clean. If the container isn't covered, dirt will get into the grease from the surrounding environment. Some companies go the extra ex-

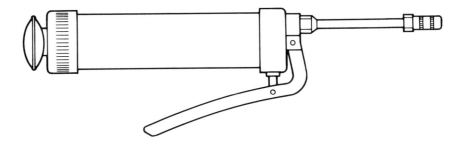

Figure 5-5. Hand grease gun.

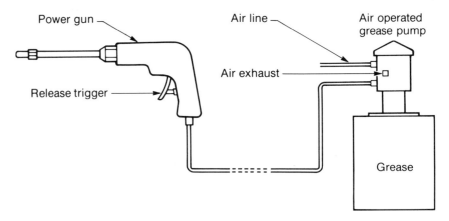

Figure 5-6. Power grease gun.

pense to use the individual cartridges of grease to insure that the grease is clean.

The *power gun* (Fig. 5-6) is used where larger quantities of grease are required. The gun is usually attached to an air-powered pump that provides the grease flow. The danger involved with a power gun is getting too much lubricant. If too much grease is used, it may churn and build up heat and actually damage the equipment it was to lubricate.

Do not point the gun at anyone. Some guns possess high pressure. If the grease hits someone at high pressure, it may penetrate the skin. This will usually result in an infection and can cause severe health problems. Safety is very important.

Automatic greasing systems (Fig. 5-7) come in a variety of styles and types. The most common has a pump attached to a

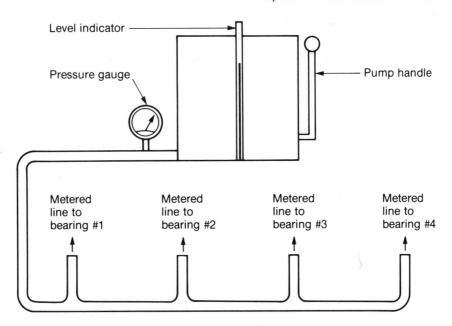

Figure 5-7. Automatic lube system.

series of grease lines. The grease lines are in turn attached to fittings at each bearing or gear. As the pump is worked, it dispenses grease in measured quantities to each point. This is a good method for it insures that each fitting will get grease. It's safer around moving machinery. The repairman can stand at a safe location and pump the system without being close to the moving machinery.

SUMMARY

Lubrication is one of the most important considerations in any mechanical drive system. The lubricant must be chosen carefully and applied judiciously. The basics are outlined in this chapter. For more detailed information, consult one of the lubrication companies.

REVIEW QUESTIONS

1. What six purposes do lubricants serve in a mechanical drive?

2. Describe the five basic lubricating systems.

3. What is meant by the term viscosity?

4. What may be the problem if a too heavy grease is used in a drive system?

5. What are the three most common methods of applying grease?

Chapter 6

Bearings

Anytime that two moving surfaces are in contact, wear occurs. To reduce the amount of friction and thus the amount of wear, bearings were developed. In this chapter we'll explore the various types of bearings used in industry today.

Friction is the problem in any drive system that has a rotating member resting on a stationary member. The problem is to find a way to reduce the amount of friction. There are two ways to solve this problem. You can use two substances that'll slide against each other with a minimum of frictional resistance, or you can introduce a rolling element to change the friction from sliding to rolling friction. These two solutions are provided by bearings. There are two types of bearings, plain or sleeve and rolling element.

PLAIN OR SLEEVE BEARINGS

Plain or sleeve bearings use low wear materials to support the load. The material usually has a very low coefficient of friction, and is usually coupled with a very good lubricant. The lubricant is used to build a wedge to eliminate any contact between the rotating parts. This may be illustrated by placing two small, flat, metal parts on top of each other. As you move them back and forth, you notice that it takes an effort to do so. Now separate the parts and place a film of oil between them and put them back together. Moving them this time takes less effort. This is the principle of plain bearings.

Plain bearings come in two basic types, hydrostatic and hydrodynamic. *Hydrodynamic bearings* develop the oil film barrier by the rotational velocity of the two rotating parts (Fig. 6-1). The rotational velocity of the bearing can build lubricant pressures of several hundred pounds per square inch. This then will lift the shaft off the inside of the bearing. The same principle applies to a car when the wheels strike a puddle of water and they hydroplane. The tire is actually lifted off the pavement.

Hydrostatic bearings depend on an external source of fluid pressure to provide the separation. They're not designed to develop a metal separating wedge.

In both classifications of bearings, there are three types of lubricating conditions:

1. **Fluid film** The surfaces are completely separated by a film of lubricant.

2. **Boundary** The surfaces are only partially separated by the film of lubricant and the rest of the load is carried by direct metal to metal contact.

3. **Extreme boundary** The surfaces are in direct contact on at least the high points or asperities. This may occur when the bearing is deprived of lubrication or under extreme overloads. This results in rapid wear of the bearing. (See Fig. 6-2.)

The sleeve or plain bearings come in a variety of materials. The following is a list of the most common types and their properties.

1. **Babbitt** A very soft bearing material, used where you don't want to damage an expensive shaft. If you have a failure the softer material will be destroyed leaving the shaft virtually undamaged. Using a babbitt bearing gives the advantage of being able to scrape a damaged bearing in the field and pouring new metal in the bearing, smoothing and putting it back into service with a very minimum of downtime. Babbitt should be

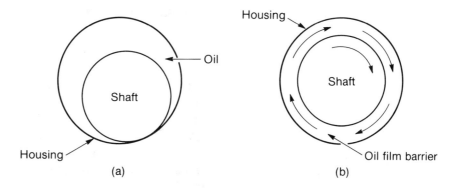

Figure 6-1. Hydrodynamic sleeve bearing: (a) stationary, (b) rotating.

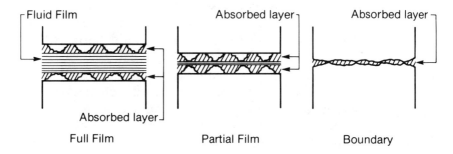

Figure 6-2. Three types of lubricating conditions.

used for light to moderate loads and operating temperatures under 200°F.

2. **Bronze** This material is used for heavier loads, slightly higher speeds, and higher temperature ranges (to 300°F) than that used with babbitt. The disadvantage of bronze is that it's a harder material and thus will have a tendency to score the shaft or housing in case of a failure. Also the bronze sleeve usually can't be repaired. Replacement of the sleeve is required.

3. **Sintered metal** This material is sponge-like in magnified surface examination. The small pores are filled with a lubricant. When rotation begins, the generated heat brings the lubricant out in a capillary action. The faults with this type of bushing are obvious. When the lubricant is used up, the bearing's life quickly ends. The advantage to this material is that it can be used in

locations that are inaccessable for bearings requiring regular lubrication.

4. **Carbon graphite** This material is used under conditions of extreme temperature (up to 700°F). The carbon graphite type of bushing is the lubricant. As the bushing wears, it lubricates. The factor that dictates when it's time to change the bearing is how much internal play can be tolerated. When the internal motion reaches the stage that it interferes with the operation of the unit, the bearing is replaced with a new one. These sleeves are usually run at light loads and low speeds.

Correct Lubricant of Sleeve Bearings

In maintenance of sleeve bearings, the correct lubricant is very important. This means not only the correct amount but also the correct viscosity. The manufacturer's recommendations for each bearing should receive careful consideration. The initial start-up temperature should be considered in selecting the lubricant. If a too high viscosity lubricant is used in an extremely cold environment, initial scoring or welding and tearing can occur. This deterioration will continue even if the lubricant reaches its correct operating temperature. If a lubricant is selected with a too low viscosity, metal to metal contact will result in quick destruction of the bearing. The point here is that in some cases lubricant changes may be seasonal. As the temperature changes, the grade of lubricant used may have to be adjusted accordingly.

Mounting Procedures

The second important point to consider in maintaining sleeve bearings is the mounting procedure. If too much mounting force is used the sleeve may be distorted, resulting in insufficient clearance in certain areas of the bearing. This pinched effect will not allow the sleeve to build up the lubricant wedge, which results in scoring and welding and quickly destroys the sleeve. A second condition that causes the same effect is a housing that's out of round. As the sleeve is pressed into or clamped by the housing it becomes egg-shaped. When the round shaft is inserted, there's no room for the lubricant to build a wedge and rapid deterioration occurs.

Lubricant Filtration

The filtration of the lubricant is important. If the lubricant is not filtered, then particles become trapped in the lubricant and are forced into the bearing. Depending on the size of the particle, as it enters the load zone it'll scratch or cut the sleeve. The scratches will accelerate the normal wear to an unacceptable level. Careful filtration is a must.

ROLLER ELEMENT BEARINGS

The advantage that rolling element bearings have over sleeve bearings can best be illustrated by a book and a table top. If you slide the book over the table, you'd have the same results as if it were a sleeve bearing (Fig. 6-3). If you were to place a few pencils under the book and move it again, you'd have the roller bearing (Fig. 6-4). If you placed a few balls under the book, it would become a ball bearing (Fig. 6-5). In each instance it takes less and less effort to move the book on the table top. The same is true with the bearings; you get less frictional resistance as you move from sleeve to roller to ball bearings.

ROLLING ELEMENT BEARINGS

Ball Bearings

In examining rolling element bearings, let's look first at ball bearings. Ball bearing types are best classified by their race configuration. They fall basically in four types:

- Deep groove
- Self-aligning
- Angular contact
- Thrust

Deep groove ball bearings. Deep groove ball bearings (Fig. 6-6) are capable of sustaining both heavy radial and thrust loads. They come in two main configurations, single row and double row. The single row is able to sustain radial and thrust loads. The double row can sustain somewhat heavier radial loads, due to the increased contact area with two rows of balls.

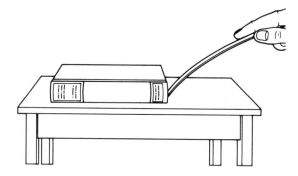

Figure 6-3. Representation of a sleeve bearing. (Courtesy of S. K. F. Industries, Inc.)

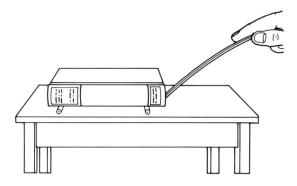

Figure 6-4. Representation of a roller bearing. (Courtesy of S. K. F. Industries, Inc.)

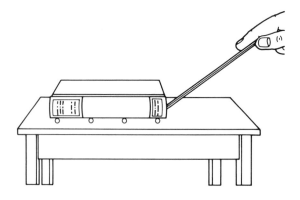

Figure 6-5. Ball bearing. (Courtesy of S. K. F. Industries, Inc.)

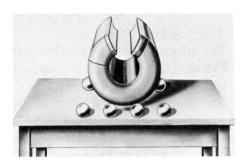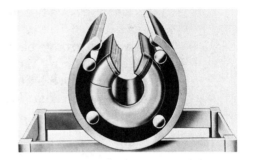

Figure 6–5. Continued

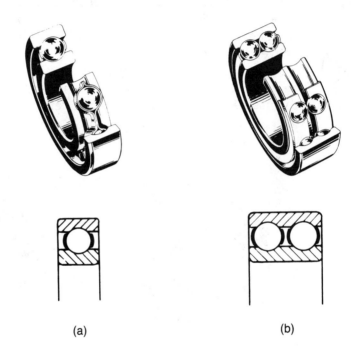

(a)   (b)

Figure 6-6. Deep groove ball bearings: (a) single row, (b) double row. (Courtesy of S. K. F. Industries, Inc.)

Deep groove bearings also come in two sub-types, conrad and max (Fig. 6-7). *Conrad* bearings have the basic configuration of any deep groove bearing. The *max* bearing has loading slots cut into the bearing races. These are cut to enable the manufacture to insert more balls into the bearing. This enables the max to carry more radial load than the conrad. However, the loading slots restrict the amount of thrust load the max bearing can sustain. If the thrust load forces the balls into the side of the race, it'll run across the loading slot. This has the effect of running your car tire through a large pothole in the street. It not only affects the tire (ball) but also the street (race). It'll rapidly remove material through impact loading until failure results.

The repairman should be alert to the type of bearing and its intended load. With this type of bearing (single row, deep groove), it's not possible to install it backward. It can be useful to begin the practice of installing this type of bearing so that the letter and number codes face outward when mounted. The advantage of this is that you can remove the bearing cap or cover, read the number, and obtain a new bearing without having to remove the old one.

need to note: which side is carrying the thrust

Loading slots

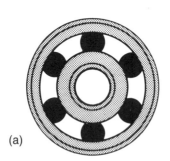 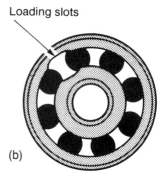

(a)   (b)

Figure 6-7. (a) Conrad and (b) maximum capacity (max) bearings. Maximum capacity have more balls.

Self-aligning bearings.   The second type of ball bearing is the self-aligning (Fig. 6-8). This bearing has the advantage of being able to adjust for some slight misalignment. The outer bearing race swivels slightly allowing the bearing to adjust for mounting errors, shaft deflections, and any base distortions. Another advantage to the curved race is the bearing race can be pivoted

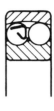

Figure 6-8. Self-aligning ball bearing. (Courtesy of S. K. F. Industries, Inc.)

out and the balls and races inspected for wear patterns or damage. The disadvantage to this bearing when compared to the deep groove bearing is that it can't accommodate much thrust load. The curvature of the outer race doesn't give the balls the necessary support to sustain any thrust loads. These bearings also come in single and double row and the amount of radial load should dictate which one is used.

Angular contact bearings. The bearings discussed so far cannot be mounted backwards; the angular contact bearing (Fig. 6-9), however, can be. The angular contact can take radial loads, but unlike the other, it can take thrust loads in one direction only. In examination of the races it becomes clear as to the reason why. At least one (possibly both depending on manufacturer) of the races are counterbored. The ball now has a small nest or track to run in. The applied thrust load should try to push the ball into this track. If the load pushes in the other direction it forces the ball to ride on the counterbored shoulder (Fig. 6-10). When this occurs you have extremely high loading in a very small area which exceeds the load capability of the bearing steel. The ball or the race will overheat and rupture the oil film barrier, resulting in the welding and tearing of the material in the balls and races. It's important to closely examine the bearing and its intended load before installing the bearing.

The angular contact ball bearing also comes in what is known as a flush-ground type. In flush grinding a bearing, the manufacturer removes material from the inner and outer rings so that

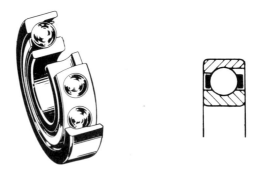

Figure 6-9. Angular contact ball bearing. (Courtesy of S. K. F. Industries, Inc.)

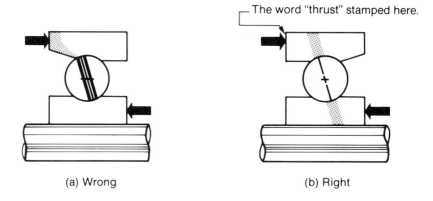

(a) Wrong   (b) Right

Figure 6-10. Right and wrong loading on an angular contact ball bearing. (Courtesy of Fafnir Bearing Div. of Textron Inc.)

when they are mounted the bearing will be preloaded. Preloading means that some internal clearances are removed from the bearing. This will reduce the amount of free movement or deflection in the bearing. This condition is required when you need bearings that'll hold a shaft rigid without any deflection. Two things to remember when preloading bearings, the higher the speed the less preload you can have. The slower the speed the more preload required. Also the more you preload a bearing the more you reduce its life.

The flush grinding of the angular contact bearings factory-sets the preload for the set of bearings. This arrangement of mounting is called *duplexing*. Duplexing can be performed in three ways: back to back, face to face, and tandem. Each of the three methods has its own advantage when it comes to mounting. When duplexing bearings the best procedure is to consult a blueprint for that particular installation. If a blueprint isn't available then it's advisable to consult your bearing distributor for the mounting procedure for your particular installation. Remember that you can use flush ground angular contact ball bearings anywhere you can use a regular angular contact bearing; however, you can't use a standard angular contact bearing in a duplex installation. Careful examination must be given anytime an angular contact bearing is being used. The flush ground bearings are plainly identified by each manufacturer on the outside of the box that the bearing is packaged in (Fig. 6-11). The markings are harder to spot on the bearing, for the identification is etched in the race. The other identification is stamped in the race. The etched marking will not show up as well as the stamped, so it'll take close inspection to find the markings.

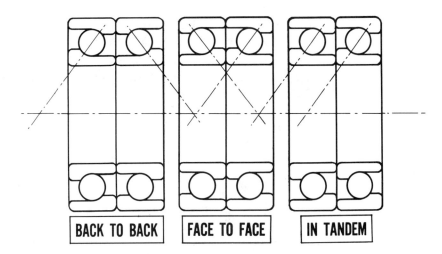

BACK TO BACK   FACE TO FACE   IN TANDEM

Figure 6-11. Coding and positioning of duplex bearings. (Courtesy of Fafnir Bearing Div. of Textron Inc.)

Thrust bearings. Thrust bearings (Fig. 6-12) are the last class of ball bearings to be considered. They're only used in cases where there are no radial loads. They're not a very common bearing but are found in some specialized installations. The basic configuration looks like a sandwich with two races and the balls in between. Any radial force exerted would cause the bearing to separate.

Figure 6-12. Spherical roller thrust bearing. (Courtesy of the Torrington Co.)

Roller Bearings

Ball bearings as a class should be used for higher speeds and lighter loads. If higher loads are encountered then some form of roller bearing should be considered. However, roller bearings do run at lower speeds than ball bearings. Roller bearings are divided into four main classes by the shape of the roller:

- Spherical
- Cylindrical
- Needle
- Tapered

Spherical roller bearings. Spherical roller bearings (Fig. 6-13) are the workhorses of the roller bearings. They're capable of sustaining very heavy radial loads and heavy thrust loads. Their spherical shape combined with the shape of the outer race gives a large contact area on which the load can be carried. The shape of the outer race also affords it compensation for some misalignment. These bearings may be found in either a single or a double row style, depending on the amount of the load. The shape of the outer race doesn't restrict the amount of thrust load that the bearing can sustain. Although the self-aligning ball bearing couldn't sustain thrust loads, the shape of the rollers allow the spherical bearing to sustain substantial thrust loading. In addition, the spherical roller bearing is designed to sustain heavy radial loads. Some manufacturers design a thrust bearing using the spherical rollers. These bearings are most frequently found in a combination style, where some radial load is also carried.

Cylindrical roller bearings. Cylindrical roller bearings (Fig.

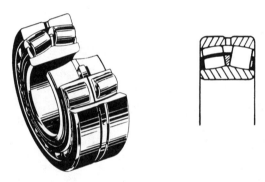

Figure 6-13. Spherical roller bearing. (Courtesy of S. K. F. Industries, Inc.)

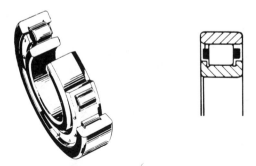

Figure 6-14. Cylindrical roller bearing. (Courtesy of S. K. F. Industries, Inc.)

6-14) are the highest speed roller bearings. They're capable of sustaining high radial loads. The bearing can't carry any thrust loads. The design is such that most makes can come apart if radial loads are applied. They may be found in single or double row types depending on the type of service.

Needle roller bearing.   If the roller's length is six times its diameter then it becomes classified as a needle roller bearing (Fig. 6-15). Needle roller bearings are unique in their design. The bearing has an outer race but no inner race. The shaft it's mounted on provides the surface that serves as the inner race. At manufacture the outer race is egg shaped. This feature allows a slight preload when the bearing is put in a round housing or is mounted on a round shaft.

One other service note on a needle bearing is the difference in the edges of the bearings. The bearing has one end that's hardened and stamped. This is the end that the mounting pressure is to be applied. The other side is spun over or rounded. If the mounting pressure is applied to the rounded edge, the rollers will be pinched or locked up. When the bearing is put in service under these conditions, it'll have flat spots worn on the rollers very quickly. If they're handled properly, needle bearings are very useful in applications with limited space.

Tapered rollers.   The last style of rollers are the tapered rollers (Fig. 6-16). The most common application of this style of bearing is in the wheel bearings on the average automobile.

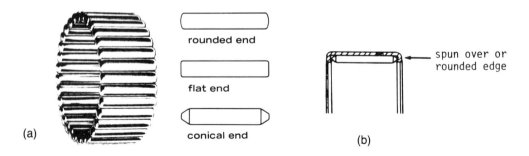

(a)

rounded end

flat end

conical end

(b)

spun over or rounded edge

Figure 6-15. Needle roller bearing. (a) Courtesy of the Torrington Co. and (b) S. K. F. Industries, Inc.)

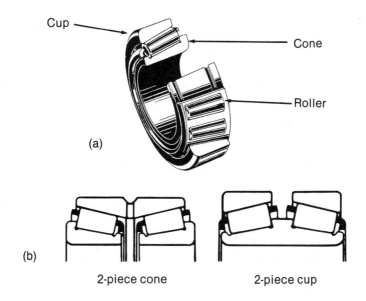

Figure 6-16. Tapered roller bearings: (a) single roller, (b) double roller. (Courtesy of S. K. F. Industries, Inc.)

These bearings carry heavy radial loads and heavy thrust loads in one direction only. If you have an application requiring the bearing to take loads in both directions, then you must use two bearings mounted opposite each other to take the loads in either direction. This type of bearing will have some type of adjustment, either on the cone or cup, to remove the internal clearances. When the bearing is in operation it can be adjusted to meet the running requirements.

MAINTENANCE HINTS

Ninety to ninety five percent of all bearings experience what is know as premature failure. This means that they didn't last as long as the manufacturer had rated their service life. Premature bearing failures can be divided into the following classes:

- Maintenance practices
- Shaft and housing fits
- Lubrication

Maintenance practices cover a large range of items that are practiced industry-wide because the average repairman doesn't know the proper procedures. One of the most common procedure is cleanliness of the bearings. The oil film wedge that's built in bearings to prevent metal-to-metal contact is from five to thirty millionths of an inch thick. The smallest particle of dirt can rupture this film barrier and result in metal-to-metal contact in the bearing. No amount of dirt, no matter how small, can be tolerated in a bearing.

When the manufacturer ships the bearing, it's clean. It comes wrapped in a acid resistant paper that protects the bearing from any outside contamination. The bearing should be kept in the wrapper until it's used. (See Fig. 6-17.) If it's unwrapped and put back in the box without rewrapping dirt will get in it. When removed and ready for use it does not need to be washed. The protective slush that is on the bearing is compatible with any petroleum based lubricants.

If it's to be used in a synthetic based lubricant then it'll need to be washed. If the bearing is washed, usually dirt will get into it then. Most equipment used to wash bearings has been used to wash everything from gears to hand tools. *Don't* wash the bearings unless you have to do so, and then only in clean fluid. Another key point in washing bearings is the fluid that you use to wash the bearing. A check should be made to see if the solvent leaves a whitish film on the bearing steel. If it does, don't use it. The deposit that the solvent leaves interferes with the proper lubrication of the bearing steel. Leaded gasolines are particularly noted for leaving this type of deposit. Kerosene is good, but it should all be removed from the bearing before the bearing is installed. One way to do this is to immerse the bearing in warm oil for a short period of time, after washing and before installation.

Anytime a bearing is handled, care should be exercised. If your bare dry hands touch the bearing steel, a chain reaction begins. The acid in your system reacts and begins a stain, which becomes an etch, then a pit, finally a complete rust spot.

If this occurs in the running area of the bearing it'll run rough and fail prematurely. You can't start a bearing in service with rust on it and expect that the rust will go away. It'll flake off, but when it does part of the bearing will go with it. When the lubricant tries to fill that area, it can't, so at the next load cycle

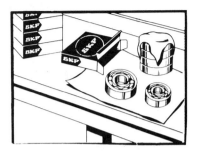

Figure 6-17. Correct unwrapping of a bearing. (Courtesy of S. K. F. Industries, Inc.)

the film barrier will be ruptured and metal-to-metal contact will occur, resulting in continued deterioration of the bearing and complete failure resulting. No contamination can be tolerated in a bearing. Once in operation, it becomes the function of the lubrication and accessory equipment to keep contamination out of the bearing, two of the more popular items being seals and shields.

Seals

Seals (Fig. 6-18) are usually made of some type of synthetic material that contacts the inner and the outer races. This will prevent the entry of any foreign material into the bearing. The bearing can be equipped with one side sealed or both sides sealed. If both sides are sealed it becomes quite obvious that relubrication is impossible. Since the bearing cannot be lubricated, its life is limited to the life of the lubricant. This then becomes a factor of the load and speed. Once the lubricant fails, the bearing will quickly follow. If you only have one side sealed, then the bearing can be lubricated through the open side. This allows for extended life of the bearing provided that the lubricant is uncontaminated.

Shields

The shields (Fig. 6-19) are inserts in the bearing that are fastened to only one race (usually the outer). There is a very small amount of clearance between it and the other race. The shield won't keep out all contamination as a seal will, but it

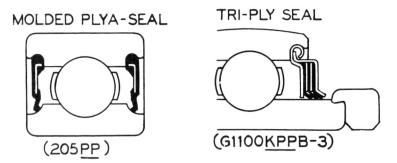

Figure 6-18. Bearing seals. (Courtesy of Fafnir Bearing Div. of Textron Inc.)

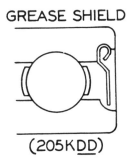

Figure 6-19. Single shield. (Courtesy of Fafnir Bearing Div. of Textron Inc.)

keeps the larger pieces of dirt, gear chips, metal shavings, etc. out of the bearing. The shield allows lubricant into the bearing, providing extended life.

Cleanliness Hints

Any method that's used to eliminate contaminants in a bearing will prevent a premature failure in a bearing. Following are basic rules.

1. Always work with bearings with clean tools and in clean surroundings.

2. When removing a bearing from its mounting, clean off all dirt before removing.

3. Always handle with care, not touching the running areas of the bearing with your clean, acidic hands.

4. Use clean solvents to wash bearings.

5. Protect clean bearings from all forms of dirt and moisture.

6. If a bearing is to be left for a while, cover it with a clear plastic. If it's to be stored, buy some of the acid resistant bearing paper to wrap the bearing for protection.

7. Always use clean, lint-free rags to wipe or dry bearings.

8. Always use clean lubricants when relubricating a bearing.

9. Always protect the bearings in storage from change in temperature. The resulting condensation can cause rust to form, ruining the bearing before it's even installed.

10. Never wash a new bearing unless you have some reason to suspect it has become contaminated. The other exception is if you're to use a synthetic based lubricant.

Item eight is a particular sore spot to most repairmen. If a grease fitting isn't wiped off before a bearing is lubricated, a small speck of dirt that collects on the fitting will be forced into the bearing. This will again cause problems, especially if it's repeated continually over an extended period of time. The grease or oil may appear to be clean, but if the lubricant sits in an open container for any period of time, dirt will be mixed in the lubricant. Anytime grease is purchased in bulk quantity and grease guns are filled from an open drum, dirt will be present. Although individual grease cartridges are more expensive, it changes the picture drastically when you factor in bearing cost.

Bearings in storage should always be protected from temperature change. Just as coming in from outside (in cold weather) to a warm room will form condensation on glasses, so will exposing bearings to temperature change in storage form condensation on the bearing. The more even the temperature stays while the bearing is in storage, the better condition the bearing will be in when it's ready to be installed.

The handling of the bearing prior to installation will greatly affect the life expectancy of the bearing. If the bearing is dropped, hit with a hammer, or spun with air, it won't last long. If you would consider a bearing as fragile as a wristwatch, it wouldn't get as much abuse. Actually, a bearing can have internal tolerances to millionths of an inch, much more precise than most watches. If more bearings were treated with the same care given a good watch, they would last much longer.

One point on air spinning bearings: don't do it. Bearings aren't made to run unsupported. If they are, they'll explode like a grenade. If you're holding it, you'll be severely injured, if not killed. It does happen. There are many recorded instances of individuals who were killed or severely maimed by spinning bearings. Some people reason, I won't spin it that fast. If the bearing is spun at all, it's usually supported on the fingers. Think of a bearing running without lubrication: it'll weld and lock up. If someone is spinning one and it locks up, the bearing will take the fingers that are supporting it off the hand.

One may ask: how will the bearing get dried if air isn't used? The only safe way is to hold both races tightly (Fig. 6-20). This will prevent the bearing from spinning, allowing it to be safely dried.

Figure 6-20. Properly air drying a bearing. (Courtesy of S. K. F. Industries, Inc.)

Shaft and Housing Fits

The second area of problems we'll consider is shaft and housing fits. This is probably the most overlooked area in bearing maintenance. Most roller bearings have some method of removing internal clearances, but with ball bearings you must depend on a shaft and a housing of the right size. If the shaft is too large in diameter you must stretch the inner ring too far to get it on the shaft (Fig. 6-21). In doing so, you remove any internal clearances in the bearing, removing the room for the metal separating wedge of lubricant to form, causing metal-to-metal contact and immediate and catastrophic bearing failure. If the shaft is too small, the inner race of the bearing will turn on the shaft causing fretting corrosion, which causes the bearing to fail prematurely.

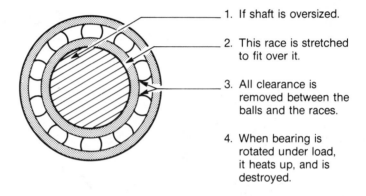

1. If shaft is oversized.

2. This race is stretched to fit over it.

3. All clearance is removed between the balls and the races.

4. When bearing is rotated under load, it heats up, and is destroyed.

Figure 6-21. Oversized shafting.

If the housing is too small (Fig. 6-22), the outer race is compressed, again removing the internal clearances from the bearing causing welding and rapid deterioration. If there is a too large clearance, then the outer race will slip again causing fretting corrosion and rapid deterioration. The only correct way to size shafts and housings is with inside and outside micrometers. The correct sizes are found in the manufacturer's specifications for each size bearing. It's most important to follow their guidelines.

Mounting of Bearings

The mounting of bearings includes more than just having shafts and housings the correct sizes. If the shaft or housing has any nicks or burrs (Fig. 6-23), this sets up additional loads the bearing wasn't designed to take. The shaft and housing should be smoothed before installing a bearing. Just because the bearing can be forced over the burr doesn't mean that it

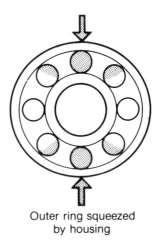

Outer ring squeezed
by housing

Figure 6-22. A bearing housing that is too small. (Courtesy of Fafnir Bearing Div. of Textron Inc.)

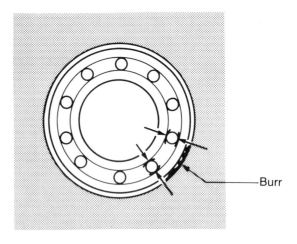

Burr

Figure 6-23. The effect of a housing burr on a bearing.

won't harm the bearing. It'll set up stresses that destroy the internal geometries of the bearing.

Mounting bearings can be accomplished by two different procedures, presses and temperature mounting.

Presses. Press fits (Fig. 6-24) are usually used on bearings less than four inches in diameter. This is merely a matter of

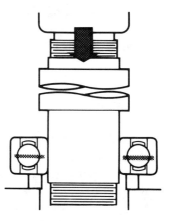

Figure 6-24. Properly pressing a bearing on a shaft. (Courtesy of Fafnir Div. of Textron Inc.)

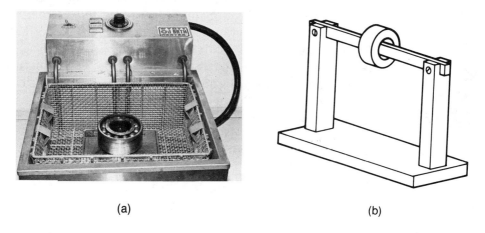

(a)   (b)

Figure 6-25. (a) Oil bath, (b) induction bar.

stretching the inner race (or compressing the outer race) till it goes over the shaft (or in the housing). This is the reason it's important to always check the size of all components in an installation.

Temperature Mounting. Temperature mounting is used on larger bearings. It's a matter of using the principle of thermal expansion to enlarge the inner race of the bearing till it slips over the shaft. This is accomplished in an oil bath or by an induction bar (Fig. 6-25). An oil bath is merely a tank of oil that the bearing is submerged in, and heated to the desired temperature. Care must be taken not to overheat the bearing. In most cases, 200°F is sufficient. No bearing should be heated over 300°F. The effect of the time and temperature will affect the bearing steel and result in premature failure.

An induction bar is a faster way to heat a bearing, and a more dangerous way. Since it is fast, a bearing can be overheated rapidly. A time chart or heat stick should be used to monitor the temperature. Most double sealed bearings are heated with an induction bar to prevent any possible damage in the oil bath. These bearings cannot be heated above 200°F. If they are, the lubricant contained inside will be destroyed. It's best not to heat a double sealed bearing above 180°F to insure some margin of safety. In temperature mounting a separable type bearing (a cone and cup tapered roller), it's possible to put the outer race in a freezer to cool it, thus shrinking it, for installation in a housing. Caution must be exercised in removing all condensed moisture before the bearing rollers come in contact with the race.

Never put a whole bearing in a freezer. The moisture that would condense inside the bearing during installation could never be removed. This will cause rust and rapid deterioration of the bearing. Another point: never play a torch on a bearing. That's too much heat concentrated in one spot. With no temperature control, the bearing steel will be damaged, shortening the life of the bearing.

SUMMARY

Bearings are the most abused and misused components in mechanical drives. They are mishandled, placed improperly in storage, installed incorrectly, and incorrectly maintained. If the guidelines in this chapter are followed, bearing life would increase dramatically. To do so means to take time to work correctly and carefully. If they're handled as carefully as a wristwatch, they'll provide longer service. Although it's difficult for the repairman to make the transition from handling heavy equipment to carefully maintaining bearings, the payoff is in not having to change the bearing as frequently.

REVIEW QUESTIONS

1. What is the basic difference between a plain bearing and a rolling element bearing?

2. What are the three basic types of lubricating conditions?

3. What are the four most common materials from which sleeve bearings are constructed?

4. What are two important considerations in maintaining sleeve bearings?

5. What are the four basic types of ball bearings?

6. What does preload do to the life of a bearing?

7. What are the four basic types of roller bearings?

8. What is the thickness of the oil film barrier that's developed in a rolling element bearing?

9. Should all new bearings be washed?

10. Describe the difference between shields and seals.

11. What can happen if a bearing is spun by air pressure?

12. What are two methods used to mount a bearing?

13. What may happen if a bearing is installed on a too large shaft?

14. What are the two recommended methods of heating a bearing?

15. Why can a torch not be used to heat a bearing?

Chapter 7

Belt Drives

Belt drives are one of the most popular methods used for power transmission in industry. There are four basic types of belts, but before we examine the different types, let's first look at some basic terms that apply to all the belts.

TENSION

Tension is the force acting lengthwise in a belt, tending to elongate it. Belt tension is usually measured in pounds. Static tension is the tension on a belt drive when it's at rest. The total tension on the drive is twice the static tension. When the drive is running, there are two types of tension. Tension in the belt, approaching the driving pulley, is the *tight side tension*. The tension in the side of the belt leaving the driving pulley traveling toward the driven pulley is the *slack side tension*. (See Fig. 7-1.) The effective tension is the tight side minus the slack side tension. If the coefficient of friction is great enough between the pulley and the belt, there'll be a large difference in tight side and slack side tension.

When you take a piece of string with a weight on the end of it and spin it quickly, you have an example of centrifugal force,

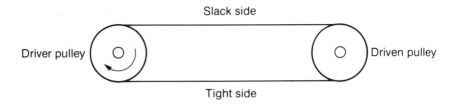

Figure 7-1. Tension zones in a pulley drive.

the tendency the weight has to travel outward in a straight line. The same force acts on a belt at higher speeds, actually trying to pull the belt away from contact with the pulley. To overcome this force, the belt has a tension factor called *centrifugal tension*. This is in addition to the tensions the belt needs to transmit power.

CREEP AND SLIP

Two other terms common in belt drives are creep and slip. Although confused with one another, they're two different occurrences in a belt drive. Slip is the result of insufficient tension on the belt, so that it actually slips on the pulley. It results in loss of speed and causes premature belt failure. Slip can be remedied by increasing tension on the belt drive.

Creep cannot be corrected for in a belt drive. Creep is the movement on the face of the pulley of the belt. When a belt passes around a pulley and there's a difference between entering and leaving tensions, there's belt creep. When the belt comes in contact with the drive pulley it's under tight side tension, so it'll be traveling the same speed as the pulley. As the belt travels around the pulley, it's under less and less tension till it reaches slack side tension. During this slackening period, the belt elements recover from elongation (or actually shorten) and move slower than the pulley face. This motion is known as creep. The creep can be reduced by increasing slack side tension, but cannot be eliminated in a belt drive. Now that we have an understanding of the necessary terms, let's look at the belt drives.

There are four basic types of belts used in power transmissions:

- Flat belts
- V-belts
- Toothed or timing belts
- Ribbed

FLAT BELTS

Flat belts are a common type of power transmission belt. In this section we're not considering conveyor belts, which are used to move material, but rather flat belts used to drive a piece of machinery. Flat belts are usually made from three materials: leather, rubber, or canvas.

Leather is the most common material and is very popular. It's usually cut from the middle portion of the hide, near the backbone. The strips of leather are usually cut in strips parallel to the backbone. The hide may vary as much as a one-sixteenth of an inch in thickness. Because of that, care must be taken by the manufacturer to insure that the belt is balanced. The belt should also be uniform in strength. The grain or hair side of the leather presents a better frictional surface than does the flesh side of the belt. The grain side is run against the pulley, and when a double belt is formed, the two flesh sides are cemented together exposing two grain sides.

When the hide is on an animal it's kept pliable by the secretions of the animal's body. When it's formed into a belt, it must be lubricated occasionally to prevent it from drying out, cracking, and breaking. This will also help maintain a good coefficient of friction between the belt and the pulley. Substances such as the belt tallow (fat), stearine (a solid fat), or fish oils are used. Mineral oils or waxes won't lubricate a belt or help preserve its coefficient of friction. Rosins will temporarily help but will eventually glaze and destroy the leather surface (also rubber and fabric). If unsure of a dressing, consult the manufacturer for its recommendation.

Rubber belts are made with a tension member, vulcanized between layers of rubber and are used in outdoor conditions or where conditions don't permit the use of a leather belt.

Canvas belts are used in any conditions, where leather or rubber belts can't be used.

Flat Belt Maintenance

There are four basic steps to flat belt maintenance:

1. **Keep belts tight.** The belts derive their power from the friction between the belt and the pulley. As the tension is increased the friction increases. The more tension the more power the belt will have. There's a limit to the amount of tension that a belt can withstand. It'll stretch with tension, and once the elastic limit of the belt is reached, more tension will not increase the amount of friction, so it won't transmit more power. If more power is still needed use a thicker leather belt, for it'll have a higher elastic limit.

2. **Keep belts clean.** If material is allowed to build up on the belts, it'll cause a lack of friction. This condition is called glazing. The material can be scraped off by using a piece of angle iron or a block of wood. Always scrape in the direction opposite the joint in the lap of the belt to prevent it from catching on the scraper and damaging the belt.

3. **Keep belts properly dressed.** Use one of the previously mentioned substances for a dressing or consult the manufacturer for recommendations. Rosins, oils, and waxes are not good dressings. The dressing should keep the belt pliable and maintain the coefficient of friction. Do not use excessive amounts of dressing, for it'll reduce the belt's stiffness and it'll have excessive give in a lateral direction.

4. **Protect the belts.** High temperature and high humidity can damage a belt. Efforts should be made to keep the belts in a cool and dry environment. All foreign substances should be kept off the belts, such as oil, grease, and dust. The drive should be protected to keep obstacles from falling into the belt or pulley. If an object is between the pulley and the belt, some of the support in the belt will probably be damaged, preventing it from tracking correctly.

Comparison of Flat Belts with V-belts

Flat belts require high tension, which means high bearing loads. V-belts require medium tension which means less bearing loads. Flat belt drives tolerate no misalignment, while V-belts can tolerate small amounts of misalignment. Neither drives have vibration. Flat belts have some belt slap, while V-belts run with no noise. Flat belts require occasional dressing; V-belts require no dressing. Neither drive requires lubrication, and the initial cost for each drive is approximately the same.

V-BELTS

V-belts derive their ability to transmit power from the contact between the sheave walls and the sides of the belt (Fig. 7-2). V-belts are never to run in contact with the bottom of the sheave. The advantages of V-belts over other types of drives are that they:

1. Permit a large speed ratio
2. Permit a compact drive design
3. Cushion the motor and bearings against load changes
4. Have no vibration or noise
5. Do not require lubrication
6. Do not shut down without warning

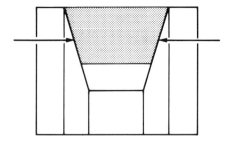

Figure 7-2. Location of tension areas of belt and pulley.

There are three main classes of V-belts: the light duty belt, the standard multiple belt, and the wedge belt.

Light Duty Belt

The light duty belt is a fractional horsepower belt (FHP) (Fig 7-3). These belts are used for light weight service and low horsepower applications. The cording and envelope of these belts are lightweight, due to the fact that the belts are generally short (10 inches to 100 inches) and bend over short radius pulleys. The approximate distance across the top of the belt is determined by the number of the belt. These belts are determined by the following number system: 2L (¼ inch across the top of the belt), 3L (⅜ inch across the top of the belt), 4L (½ inch across the top of the belt), 5L (⅝ inch or ¾ inch across the top). The length code for the belts is as follows: 4L 150 belt is a 4L (½) belt, 15.0 inch long, outside length.

Standard Multiple Belt

The next class of belt is the standard multiple belt (Fig 7-4). It's available in five different cross sectional areas. They're approximately A (⅛ inch or ½), B (⅝ inch or ¾ inch) C (⅞ inch), D (1⅛ inch or 1¼ inch) and E (1⅖ inch or 1½ inch). The distances given are the widths across the top of the belt. These belts are available in sizes from 25 to 660 inches. They have multiple cord construction for added strength and shock loads. The length code is as follows: A B-70 means a B cross section V-belt with a length of 70 inches.

Wedge Belt

The last type of V-belt is the wedge belt (Fig. 7-5). They are an improved design V-belt. These belts can transmit equal power with a smaller cross sectional area. They use three sizes to replace the five standard sizes: 3V (⅜ inch across the top), 5V (⅝ inch across the top), and 8V (1 inch across the top). The belts are run with higher tensions than the standard belts, and usually have to be tensioned with a tensioning tool, which will be discussed later. Lengths on this belt run from 25 inches to 500 inches. The coding system is identical to the FHP belts. For example, 3V 250 means you get a 3V cross section belt 25.0 inches long.

V-belt Installation

The first thing to check when installing a set of V-belts is the alignment of the sheaves. There are three types of misalignment that can affect a belt's life:

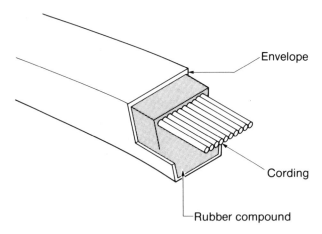

Figure 7-3. Fractional horsepower (FHP) belt.

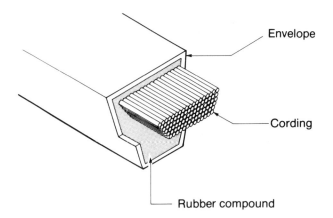

Figure 7-4. Standard multiple belt.

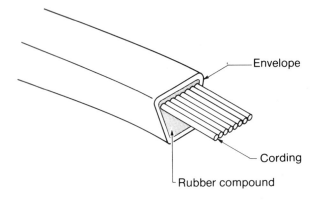

Figure 7-5. Wedge belt.

- Angular (Fig. 7-6a)
- Parallel (Fig. 7-6b)
- Sheave misalignment (Fig. 7-7)

There are ways to check alignment of the pulleys before installing new belts that'll be convenient for the repairman. The first way is to stretch a piece of string across the face of the pulleys, making sure it touches on all four points (Fig. 7-8).

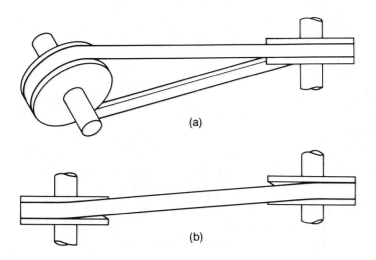

Figure 7-6. Angular (a) and parallel (b) alignment of sheaves.

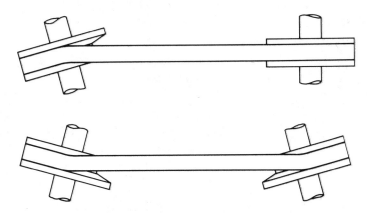

Figure 7-7. Sheave misalignment.

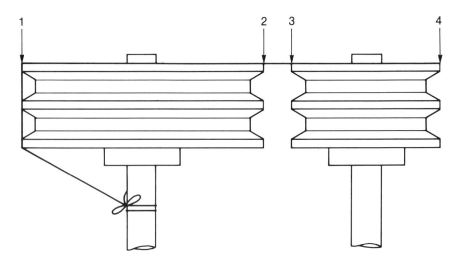

Figure 7-8. Checking pulley alignment.

The second way is to use a piece of string and two squares to make sure the shafts are aligned.

The third way is to use a machinist level and measuring bars or steel rule.

Follow these steps to install V-belts:

1. Reduce the center-to-center distance of the pulleys. Belts are not to be pried into the sheaves. This'll break the tension members (cords) in the belt. It won't carry its full load with the cords broken. This'll also cause the belts to turn over in the sheave grooves, or to narrow down in the place where the cords are broken.

2. Keep the belt slack on one side when tightening the belts. This'll insure that all belts will be carrying the same load.

3. Tighten the belts. This step is important because all belts must be properly tensioned for maximum service life. Too much tension puts too great a load on bearings and drive components. Too little tension causes the belts to slip, and thus damages the belts and sheaves.

4. Recheck belt tension after a run-in period of 24 to 48 hours. This'll insure that the initial belt stretch has not left the belt too loose to provide proper load carrying capacity.

After these steps are completed, the belts and sheaves will provide you with the maximum service life. The first three steps mentioned will not provide maximum life if step four is not properly carried out. The tensioning of a belt is the hardest step in installing new belts. Most repairmen feel as long as the belt isn't squealing that it's tight enough. This isn't true. A V-belt can slip up without squealing. You can lose up to one fifth of your belt capacity without realizing it.

Standard multiple and the fractional HP belts, when properly tensioned, will have a spring when struck by an object. They'll have a dead feel to them if they're too loose when struck by your hand. If they're too tight they will feel as if they have no give to them; they will be rigid.

The correct way to tension them is with a belt tensioning tool that may be purchased from any belt distributor. Wedge belts can't be tensioned by hand, because they must be so tight. Most repairmen don't give them enough tension. The only correct way to tension a wedge belt is by the use of a tension tool.

To protect a V-belt drive, one of the first items to check is the sheave groove. Worn sheave grooves (Fig. 7-9) will shorten belt life. If the sheave grooves are worn, the belts may bottom out in the grooves. This'll allow the belt to slip excessively. The sheaves usually dish out on the sides when worn. When this happens the sheave will wear the corners of the belt. Continued slippage and eventual destruction of the belt will result. A sheave groove gauge can be used to determine excessive wear (Fig. 7-10). The gauge can be placed in the sheave and a feeler gauge used to measure the wear. If there's excessive clearance, the solution is to replace the sheaves.

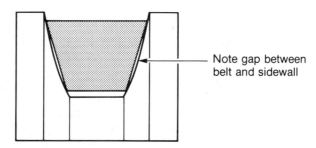

Note gap between belt and sidewall

Figure 7-9. Installation of a new belt in a worn sheave groove.

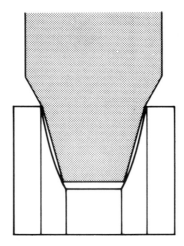

Figure 7-10. Sheave groove gauge.

Replacement Belts

The next point of consideration is the proper replacement belt. When replacing belts on a multiple drive, don't mix old and new belts (Fig. 7-11). The old belt will be elongated, and will be longer than the new replacement belt. The old belt, then, will not carry its share of the load. The new belt will be overloaded and will fail prematurely. All new belts should be checked to be sure that the match code is the same. The code is as follows:

C 40 (manufacturer's name) 50

The C 40 tells you that the belt is a C cross section, standard multiple belt with an outside length of 40 inches. The match code 50 tells you that the belt is 40 inches long. If the code is more than 50 (for example 51, 52, 53) it's longer than its stated length. The more over its stated length, the higher the match code number. If the number is less than 50, then the belt is shorter than its stated length. The less the number (49, 48, 47, 46, etc.), the shorter than its stated length. The belts should be checked to insure that they're made by the same manufacturer. The belts, even with the same match code, vary somewhat in length from manufacturer.

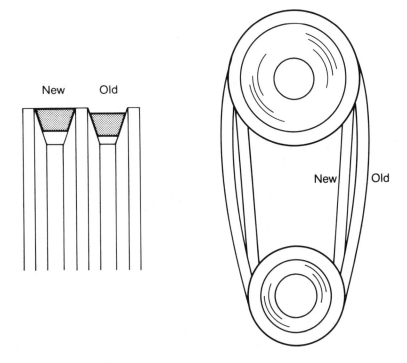

Figure 7-11. New and old belts.

As the belts are installed they should never be pried on the sheaves. Another bad installation practice is to run the belts on the sheave. These methods of installation will damage the internal cords and render the belt unfit for service.

Belt Tension

The next point to consider is the belt tension, for this can dramatically affect belt life. The ideal tension is the lowest tension that the belt won't slip under peak load conditions. Over tensioning shortens both belt and bearing life. The sheave alignment must be maintained while tensioning the drive. The belts must be kept free from any foreign material which may cause them to slip. Belt tension should be checked frequently during the first 24 to 48 hours of run in operation. After the initial tensioning period, the belts should be checked on a periodic basis for proper tension.

Guards

The guards should always be replaced to prevent foreign objects from falling into the drive. Guards should be made of screened or meshed material because they allow air to circulate, preventing the build up of excessive heat.

Idlers

If a drive has an idler it should be properly adjusted during installation of the belts. A properly designed V-belt drive won't require an idler if proper tension can be maintained. The use of idlers should be avoided, if at all possible. The idler will put an additional bending stress point on the belt, which reduces its service life.

The inside idler is located on the inside of the belt on the slack side of the drive (Fig. 7-12). The idler should be located near the large sheave in the drive, to avoid reducing the arc of contact on the small sheave. The size of the idler pulley should be at least equal to or greater than the size of the small sheave.

A back side idler increases the arc of contact on both sheaves; however, it also forces a back bend in the V-belt that contributes to premature failure (Fig. 7-13). The idler also puts extra stress in the bottom of the belt, which will result in the bottom cracking. This idler shouldn't be used unless absolutely necessary. If it's used the pulley should be at least $1\frac{1}{2}$ times the size of the small sheave and located as close to the small sheave as possible.

The kiss idler differs from either of the other two, for it doesn't affect the belt span (Fig. 7-14). Since it doesn't bend the belt, it won't contribute to premature belt failure. It isn't a common idler, but it's used for controlling vibration and whip on drives with shock and pulsating loads. If this idler is used, the diameter of the pulley should be $1\frac{1}{2}$ times the diameter of the small sheave.

V-Belt Maintenance

Since V-belts are basically a trouble-free drive system, they usually don't receive the minimal care needed to deliver maximum service life. The belt runs till it is destroyed and then a new one is put on and the drive restarted. What if the repairman could look at the belt and determine what caused the failure,

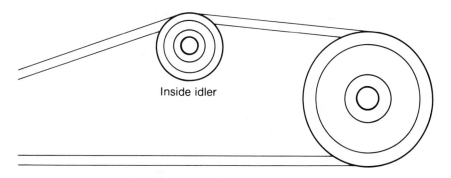

Figure 7-12. Inside idler.

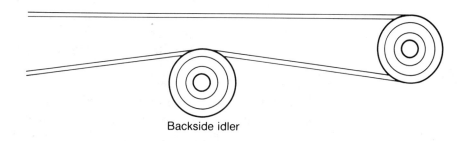

Figure 7-13. Backside idler.

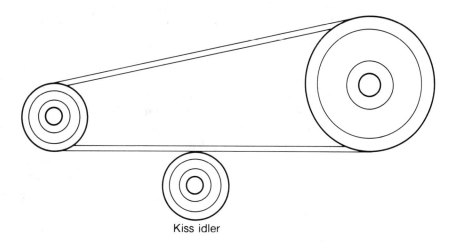

Figure 7-14. Kiss idler.

before installing new belts? This would allow him to correct the problem and then get maximum wear out of the new belts.

Belts exposed to oil and grease will fail prematurely. Any oil or grease leakage on the belts should be prevented. Repairs should be made to stop the leak. If the leak cannot be stopped, special oil-resistant belts should be used. The other problem could be too little oil and grease in the bearings, resulting in high loads and premature belt failure. The sheaves should always be checked for free rotation before installing new belts. Dirt and dust will rapidly accelerate belt envelope wear. When the belt enters the sheave, the dust and dirt get between the belt and the wall of the sheave. The belt will then slip because of not having good frictional contact with the pulley. The slippage combined with the dirt and dust in the sheave will leave the belt envelope in the same condition as if you rubbed it with sandpaper. The envelope wear will progress till complete belt failure results.

Belt guards are mentioned again here only as a reminder. They should be vented to prevent heat build-up and also should be checked to insure they're not rubbing the belt. If the guard rubs against the belt, it'll damage the envelope and result in premature belt failure.

Belt dressing should never be applied to any V-belt. The system is designed to run without belt dressing. Dressing will only contribute to belt wear and eventual failure.

The cracks that a belt develops in the bottom envelope and rubber won't reduce the tensile strength or the operating efficiency of the belt. The cracks are caused by high temperatures, small diameter pulleys, or a back side idler. It isn't necessary to replace a belt simply because bottom cracking has been observed.

V-belts are cured with a controlled heat for a given time. Belts can operate at a temperature of 140°F or less, without being affected by the heat. At any temperature greater than 140°F, the belts will overcure and not give maximum service life. Environmental conditions should be considered when installing V-belts, for excessive temperatures may indicate the need for special belts.

Causes and Cures of V-belt Problems

The following section describes belt wear and maintenance problems. The probable causes and cures are described.

Rapid belt wear. On a belt drive which shows rapid belt wear, there are several items to check.

1. Check to see if the guard is rubbing the belt.
2. Check to see if the sheaves are worn.
3. Make sure the belts are matched in a multiple belt drive.
4. Check the sheaves for proper alignment.
5. Check the tension to insure there's no belt slippage.
6. Make sure the drive is free of dust or abrasive contaminants.
7. Make sure the belts are properly installed.

Belts turning over in the sheaves. When the drive is started and run for a short period of time, the belts may have a tendency to turn over in the sheaves. If this condition exists, check the following:

1. Check the drive alignment.
2. Check the tension. If the drive has pulsating loads, the tension may need to be increased.
3. The belts may have broken internal cords. The only correction here is to properly install new belts.

Belts slipping slightly. When the drive is running and you notice the belts slipping slightly, check the following:

1. Proper belt tension.
2. Check to see if the belts are bottoming in the sheave.
3. See if there's oil or grease leaking on the drive.

Excessive belt slippage and belt squeal. When the slippage becomes excessive (over 20%), it'll be accompanied by belt squeal. This can be corrected by the same steps in the previous paragraph, with the following exceptions:

1. An overloaded belt drive may require the addition of another belt to the drive, or a higher rated belt.

2. A too small sheave won't carry the load. Check to see if any recent changes were made in the drive system. The size of the sheave or belt may need to be increased.

Belts breaking. If a belt drive has a continual problem of breaking belts, the following items should be checked:

1. See if the belt drive is subject to shock loads. If so, then the belt tension needs to be increased.

2. Check to see that proper installation procedures are being followed. It's possible the belts are suffering damage during installation.

3. Check to insure there's no slack in the belt. If there is, apply proper tension. A slack belt subjected to a sudden tension will have incalculable stresses put upon it.

Loose cover or envelope. If a drive is inspected and the belt has a loose cover or envelope, or is swollen, the following should be checked:

1. Is the belt properly ventilated so that the heat build up isn't excessive? Is it installed in an excessively hot area? If so, provide a method of ventilation or consult the belt distributor for a specially designed belt.

2. The drive should also be checked to insure that the drive isn't getting oil or grease on the belt. If so the condition should be corrected and the belts and sheaves cleaned or replaced.

Hardened or cracked belt. If the underside of the belt becomes hardened or cracked, the following items should be given attention:

1. If the heat is excessive, the belt should be properly ventilated or a special-construction, heat-resistant belt should be used.

2. Belt slippage will also cause excessive frictional heat between the belt and the pulley. This'll cause hardening and cracking. Retensioning will cure the problem.

3. The use of a substandard back side idler or a too small sheave will cause this also. Drive design should be checked.

Stretched belt. If a belt appears to be stretched, consider the following:

1. In the first 24 to 48 hours of operation some initial stretch and seating is normal. Readjustment of the tension is all that's required.

2. If the stretch continues to be a problem, the belts will most likely have broken internal cords. Replacing the belts, following proper installation procedures, is recommended.

Narrow spots on belt. If the belt develops narrow spots in it, the following is to be considered:

1. Internal cords are broken due to the shock loads or poor installation practices. Run the belts to destruction and properly install new ones.

Whip. If the belts have considerable amounts of whip, check the following:

1. Check the tension in the belt. If the belt has slack in it, it'll have whip. Tension it properly.

2. The center-to-center pulley distances will be too long. Install a kiss idler to dampen the vibration.

If a belt develops a chirping sound, it's nothing serious. The chirping will sound like a bird. Dust is usually the problem. Never try to correct this problem by applying oil or dressing. Realignment of the sheaves and idler may help. Squeak is annoying, but it won't harm the belts.

TIMING BELTS

The timing belts can be used in more different types of applications than any other belts or drive systems. They're economical and efficient, and provide superior performance.

Timing belts can transmit many hundreds of horsepower and can operate in speed ranges of 0 to over 16,000 F.P.M. (feet per minute).

In comparision with other power transmission systems, the timing belt drives have outstanding advantages. They don't rely on friction because of their positive slip-proof engagement (Fig. 7-15). They don't require high tension as do the V-belt and flat belt drives. This produces less loading of the bearings and related drive equipment. Timing belts run free of chatter and vibration. The belts are designed so that there's an extremely small amount of backlash. They frequently outperform gear drives. The fact that they require no lubrication and low tension makes the timing belt drive virtually maintenance free. The teeth of the belt are designed to provide positive engagement with the mating grooves of the pulley. The teeth enter and leave the pulley with a rolling action, with very little friction. The belt teeth function the same as the teeth on a gear. The belts don't rely on thickness to develop their strength. They can be thin to reduce heat build-up without sacrificing strength. The belt construction consists of four parts. See Fig. 7-16.

1. **The tension member.** This is a continuously wound tension member. It has a high tensile strength, excellent flexibility, and resistance to elongation.

2. **The neoprene backing.** The backing and the teeth are made from the same material. They're molded together during the construction of the belt. The flexible neoprene protects the tension members from oil and moisture.

3. **The neoprene teeth.** The teeth are precisely molded and accurately spaced. They are designed so the pitch of the teeth is not changed when the belt is flexed. The tooth shear strength exceeds the strength of the belt when six or more teeth are engaged in the pulley.

4. **The nylon facing.** This is a fabric facing that covers the toothed side of the belt. It's designed as a protection for the toothed surfaces. It normally outlasts the other components of the belt.

Most manufacturers produce these belts with five different pitches: X1 = $\frac{1}{5}$ inch, L = $\frac{3}{8}$ inch, H = $\frac{1}{2}$ inch, XH = $\frac{7}{8}$ inch, XXH = $1\frac{1}{4}$ inch. The belts will be specified by a code. They appear as follows: 225 L 075. The 225 is the length multiplied by 10. The L is the pitch code (L = $\frac{3}{8}$ inch). The 075 is the width divided by 10; for example, $\frac{3}{4}$ inch = .75 divided by 10 = 075.

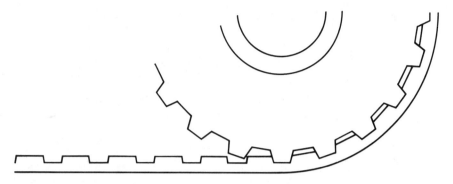

Figure 7-15. Timing belt drive.

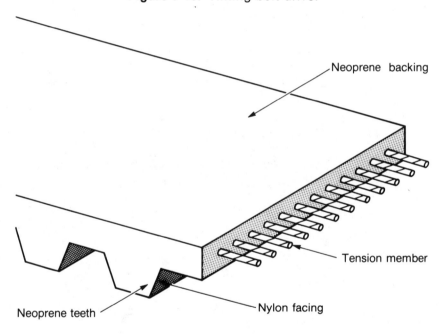

Neoprene backing

Tension member

Neoprene teeth

Nylon facing

Figure 7-16. Construction of a timing belt.

Installing Timing Belts

When installing timing belts, proper tension is important. The belt's positive grip eliminates the need for high tension. The belt should be installed with a snug fit. This'll produce less wear on the bearings and provide quieter operation. When the belt is installed with excessive tension, it'll fail prematurely. If the belt is too loose, it'll jump teeth under load, which will shear off the teeth.

The belt should never be pried over the sheave during installation. Reducing the center-to-center distance or the idler tension will permit easy installation. If this cannot be done, remove the pulleys, place the belt on them, then reinstall the pulleys.

Drive misalignment results in unequal tension and belt edge wear. Pulley alignment should be checked to assure that they're parallel. It's also important on long drives to have enough tension on the belt to prevent the tight side teeth and the slack side teeth from coming in contact with one another.

Idlers can be used if necessary with timing belts. The idler should be on the slack side of the belt. It's recommended that the idler pulley be grooved. The idler shouldn't be crowned, and should have flanges to insure proper belt tracking. The idler diameter should always be larger than the smallest recommended pulley.

Causes and Cures of Timing Belt Problems

The following section describes belt wear and maintenance problems. The probable cause and cures are described.

Excessive edge wear.   If the belt experiences excessive edge wear, the following points should be checked:

1. Misalignment. Check to insure that the pulleys are correctly aligned.

2. Damaged flange. Check to see if the flanges are damaged. If they are, repair or replace them.

Backing problems.   If the belt develops problems with the backing, consider the following points:

1. Environmental temperature. If the temperature is too low, the backing will develop cracks. If the temperature is too high, the back will soften. If the conditions cannot be changed, consult your distributor for a temperature resistant belt.

2. Oil or grease. If oil or grease gets on the belt it will soften the neoprene backing. Repair leak or use an oil resistant belt.

Teeth shearing off.   If the teeth shear off of the belt, consider the following:

1. Is the pulley diameter too small? If there are less than six teeth in mesh, tooth shear can occur.

2. Is the belt load above its rated load? Is there a mechanical bind in the drive resulting in an overload?

Teeth wear.   If the belt wears rapidly on the toothed side of the belt look for the following:

1. Does a drive overload condition exist?
2. Tension. Is there excessive tension on the belt drive?

RIBBED BELTS

Ribbed belts are single units with a longitudinally ribbed traction surface (Fig. 7-17). The ribs mate with sheave grooves of the same shape. The ribbed belts combine the fine qualities of both the flat belt and the V-belt. Ribbed belts have a greater area of belt and sheave groove contact than either V-belts or flat belts. There is less wear on the belts or sheaves than on the other two styles of belts. The sheaves are lighter and smaller, which means more compact drives. Speed ratios as high as 40:1

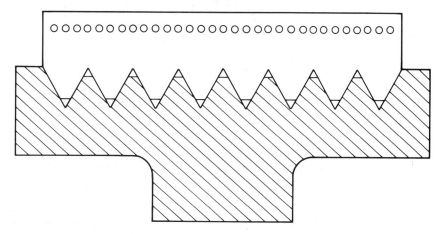

Figure 7-17. Cross section of a ribbed belt.

are possible. There's no matching problem and the belts cannot turn over in the sheaves.

The belts are composed of five basic parts (Fig. 7-18). The tension member is comprised of specially treated cords that provide stability and long flex life. The cords are then sealed in a special oil and heat resistant material. The rubber ribs give support to the cords and have a high wear resistance. The ribs are coated with a protective synthetic facing. This allows for maximum wear resistance and protects the ribs from cracking. The backing is oil and abrasion resistant to give the belt cords protection from the environment.

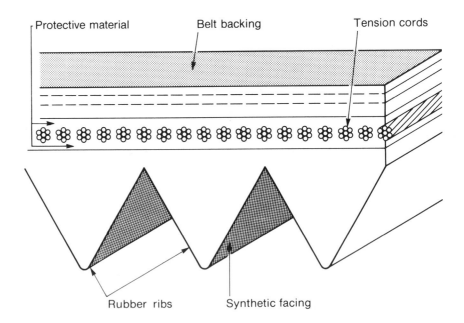

Protective material   Belt backing   Tension cords

Rubber ribs   Synthetic facing

Figure 7-18. Construction of a ribbed belt.

The belts come in five cross sections, H, J, K, L, M, being the most common. The belt code is as follows: 1120 L 16. From that number you would get an "L" section belt. It would have a pitch length of 112.0 inches. The belt would have sixteen ribs on it.

As with the other belts, the alignment of the sheaves is important. They can be aligned in the same manner as described in the section on V-belts. When tensioning the ribbed belts the manufacturer's recommendations should be obtained from the belt distributor and these should be followed. The care and troubleshooting of these will be the same as for the other style belts.

SUMMARY

Belt drives are common in industry today. Repairmen work on these drive systems frequently. The proper care and maintenance of these systems is important to the life of the drives. Proper installation of these belt drives is very important. The drives require little maintenance. A repairman should be able to identify any drive problems by merely looking at the drive or listening to it. Once a problem is spotted, it should be corrected before damage occurs to the belt.

Belt drives come in a large variety of sizes and shapes. At least one type of belt can be used in almost any drive condition. Once proper selection is made and proper care given, the drive will have a worry-free service life.

QUESTIONS

1. List and describe the four main types of belt drives.
2. Explain the difference between belt slip and belt creep.
3. List the three types of flat belts and the types of conditions under which each is used.
4. Describe the four steps of flat belt maintenance.
5. List the comparisons between flat and V-belts.
6. What are the three most common types of V-belts?
7. Describe the proper installation procedure for V-belts.
8. How can excessive sheave wear be determined?
9. In the code C 40 _____ 50, what does the 50 represent?
10. Describe the three types of belt idlers.
11. How does dust damage a belt?
12. List four causes of rapid belt wear.
13. List two causes of frequent belt breakage.
14. Why do the timing or gear belts not require high tension?
15. What effect does misalignment have on a timing belt?
16. Describe several advantages of ribbed belt drives.

Chapter 8

Chain Drives

V-belts are used to transmit power between components with long center distances. V-belts are not positive drives. Some slippage may occur under heavy loads. If the slippage is objectionable, a roller chain drive may have to be used.

Roller chain is a flexible and positive method of transmitting power between shafts at long center distances. The chain can be very efficient if it's kept well lubricated.

ROLLER CHAIN LINKS

A standard roller chain is composed of two basic links: the *pin* and *roller* (Fig. 8-1). The pin is pressed into the linkplates and does not rotate. The bushing is pressed into the roller linkplate. The pin and roller joint provide the pivoting action necessary for the chain to travel around the sprocket. The roller is free to rotate around the bushing. The roller must not be held by the roller linkplate. There must also be clearances between the roller linkplate and the pin linkplate to provide the freedom to pivot. When the chain is lubricated the lubricant must be thin enough to penetrate between all of the above mentioned areas.

Lubrication is the most important maintenance consideration in dealing with roller chain. (See Fig. 8-2.) Roller chain wears

C-1998

C-1994A

Figure 8-1. Pin and roller links. (Courtesy of P. T. Components Inc.)

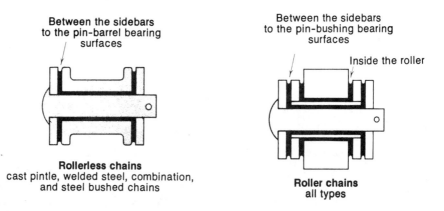

Between the sidebars
to the pin-barrel bearing
surfaces

Rollerless chains
cast pintle, welded steel, combination,
and steel bushed chains

Between the sidebars
to the pin-bushing bearing
surfaces

Inside the roller

Roller chains
all types

Figure 8-2. Areas of lubricant penetration. (Courtesy of P. T. Components Inc.)

three-hundred times faster unlubricated than it does when it's lubricated. If a roller chain was designed to last a service life of 7 years, and is run unlubricated it would have an estimated life of 9 days. Certainly this shows the importance of proper lubrication.

Two other links that may be seen in a roller chain installation are the *connecting* and the *offset* link (Fig. 8-3).

The connecting link is used to make a roller chain endless. It looks like a standard pin link except that one of the link plates will not be pressed on the pins. It will have slightly larger holes cut into the plates. This allows the link plate to be slipped over the pins. Some means will be provided to prevent the link plate from coming off the pins, usually cotter pins or a spring clip.

Two types of connecting links are presently manufactured.

C-1996

C-1996

C-1995A

Figure 8-3. Connecting and offset links. (Courtesy of P. T. Components Inc.)

With one type the link plate slips over the pins easily; the other type requires a slight press force to put the link plate over the pins. Both types will, however, still have some retainer to keep the link plate in position. One other point to keep in mind when installing a connecting link is to push the link plate out against the retainer once the installation is complete. This allows the lubricant to penetrate the area between the linkplate and to get into the roller-bushing-pin joints.

The offset link is the other link used in roller chain. The offset link is also used to connect roller chain. It will however leave the chain with an odd number of pitches. This link can be used to adjust the length of a chain when the center-to-center distance of the sprockets can't be adjusted. It shouldn't be used unless there's no other option because it isn't as strong as a standard roller chain link.

All of the four standard roller chain links are case hardened. This means that they have a hardened outer surface, but the inner surface is softer material. When the chain is worn enough that the hardened surface is worn away it should be replaced. This'll usually be about a 3% wear. If you have a length of roller chain 100 inches long, and after removing it from service it is 103 or more inches long, it's worn to the state that it must be replaced. Failure to do so will result in a broken chain during operation and subsequent loss of production.

ROLLER CHAIN SIZING CODE

The standard roller chain sizing code is important for the repairman to know. (See Table 8-1.) It's standard in industry and is simpler than the belt coding system. The length code is divided into the eighths system. The first digit (or in the case of larger chain, the first two digits) indicates the pitch length in eighths of an inch. If you have a 40 chain the pitch length is $\frac{4}{8}$ or $\frac{1}{2}$ inch. The zero indicates that it's standard roller chain. A 60 chain would have a pitch of $\frac{3}{4}$ inch and standard roller chain.

A 25 chain would have a pitch of $\frac{2}{8}$ inch but the 5 indicates that it's rollerless chain. The roller is left off because of the small size, and the bushing has direct contact with the sprocket. This style of chain is usually found in the small sizes only (25, 35 being the most popular). A 41 chain has a pitch of $\frac{4}{8}$ inch and the 1 indicates that the chain is light chain, meaning it's narrow and for lighter loads. A 60-H chain is a $\frac{3}{4}$ inch pitch chain with a heavy series designation. This means that the side plates are

Table 8-1   Chain Sizing Code

Chain Number	Pitch (in inches)	Type of Roller Chain
25	$\frac{1}{4}$	Rollerless links
30	$\frac{3}{8}$	Standard links
35	$\frac{3}{8}$	Rollerless links
40	$\frac{1}{2}$	Standard links
41	$\frac{1}{2}$	Light duty, narrow links
50	$\frac{5}{8}$	Standard links
60	$\frac{3}{4}$	Standard links
60-H	$\frac{3}{4}$	Heavy series links
80	1	Standard links
100	$1\frac{1}{4}$	Standard links
120	$1\frac{1}{2}$	Standard links
140	$1\frac{3}{4}$	Standard links
160	2	Standard links
180	$2\frac{1}{4}$	Standard links
200	$2\frac{1}{2}$	Standard links
240	3	Standard links

Note: Any chain sized above the 60 series may have the -H designation.
Note: Any chain having a -2, -3, -4H etc. is designated as having that number of strands.

thicker than the normal chain. In fact a 60-H chain has the same size (thickness) link plate as an 80 standard series chain. The heavy series chain doesn't start until the size 60. From 60 on to the larger sizes, this designation may be found. One last variable in the code is a number designation following the code. For example, an 80-3 is an ⅝ or 1 inch pitch chain that's three strands wide. The purpose of having more than one strand is to increase the horsepower rating of the chain.

INSTALLATION OF ROLLER CHAIN

For proper installation of a roller chain you should follow these steps:

1. Remove the old chain and clean the sprockets.
2. Check the alignment of the sprockets.
3. Check the sprocket for excessive wear. If the teeth become hooked shape, they'll damage a new chain.
4. Lay the new chain over the sprockets and connect.
5. Increase the sprocket's center-to-center distance to tension correctly (approximately 2% sag at the center of the strand).
6. Check for proper lubrication.
7. Start drive and recheck tension after 24 hours of run time. Retension if necessary.

ROLLER CHAIN WEAR

When a roller chain wears, the pin and bushing joint clearances increase making it appear that the chain has stretched. Since only the outer surfaces are hardened (Fig. 8-4), once they are worn away the softer inner core must bear the load. Since it's soft, it wears at a quick pace and chain failure results very shortly. As the wear progresses and the chain does lengthen it begins to ride higher on the sprocket teeth. This increases the loading forces on the chain and sprocket. As a general rule a 3% elongation is all that can be tolerated before the chain should be replaced. At this point most of the hardened surfaces on the chain bushing and joint can be considered to be worn away. While this is good general rule, sprockets having large numbers of teeth can tolerate slightly more elongation. The best way to

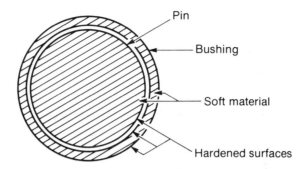

Figure 8-4. Enlarged pin and bushing joint.

measure the elongation is to remove the chain and lay it on a flat clean surface. If you have a 100-inch section of chain and it measures 103 inches, replacement should be made.

TYPES OF CHAIN FAILURE

When a chain failure does occur, it's important to inspect the chain to find what caused the failure. Roller chain failures usually can be classified in three categories: wear, fatigue, and ultimate strength.

Wear

Wear can be broken into two parts, chain or sprocket. As previously stated, as a chain wears it appears to elongate. This elongation between the links causes it to ride high on the sprocket teeth. When the links go through the slack side of the chain, there will eventually be enough slack chain for it to jump a tooth. This'll cause shock loading and accelerate chain wear, possibly even breaking the chain. The most common cause for this wear is the lack of the proper lubricant. When metal to metal contact in the pin and bushing joint occurs, metal is quickly removed. With this hardened outer material removed, the softer material wears even faster. Premature failure quickly follows. One other way the wear may occur is when abrasive material gets on the chain and works itself into the chain joint. This is like taking sandpaper to the chain, and rapidly removing chain metal, resulting in the elongation of the pin and bushing area.

Sprocket wear is a normal occurrence in any drive. The contact between the chain and sprocket teeth causes eventual metal removal even under the best lubricating conditions. The sprocket doesn't have to be changed every time the chain is replaced, for a sprocket may outlast three or more replacement chains. However, when chain replacement has become excessive, thought should be given to possible sprocket replacement. The worn sprocket will accelerate chain wear to a fantastic degree due to the loading that occurs when the chain isn't riding in the designated area on the sprocket tooth.

Fatigue

Fatigue failure can occur in any component of the chain. The link plate will almost always experience a fatigue failure near the hole in the plate (either the pin or the bushing). The crack that develops is due to an overload that exceeds the rating of the link plate but does not exceed its ultimate strength. This repeated stressing causes the crack to expand and finally breaks the link plate. The loading of the chain should be checked to insure that the manufacturer's recommendations aren't exceeded.

Bushing fatigue failures occur in a drive system due to continuous overloads. The cracks from the overstressing of the bushing form in two ways. The first way is a crack running the length of the bushing, usually caused by too hard an impact of the chain with the sprocket teeth. The second way is a circumference crack around the bushing at the link plate. Either type can be prevented by reducing the loading on the chain to a level that doesn't exceed the manufacturer's recommendation for the chain.

Roller fatigue failures are caused by the impact of the chain on the sprocket teeth. This may also cause bushing failure due to the transfer of load from the roller to the bushing. However, the roller isn't affected by the load as much as the bushing. The roller is affected by the speed of the drive. The faster the chain runs, the more shock the roller must absorb as the two come together. The roller may also absorb high loads if the chain is jumping teeth on the sprocket. The constant over-loading causes the roller to fatigue. The cures then become obvious, reduce the speed for the size chain, or correct the problem causing the chain to jump teeth.

Pin fatigue failure rarely occurs in a chain drive, for the pin has a fatigue strength far in excess of the other chain parts. Sometimes an overload will occur beginning a crack in the pin, and progressing to a failure that appears to be a fatigue failure. When this occurs it's virtually impossible to detect the difference. Using a higher rated chain or reducing the load is the only cure for this condition.

Ultimate Strength Failures

Ultimate strength failures occur in link plates, bushings, and pins from a large overload being applied as a sudden shock to the drive. This may occur as a sudden shock or a normal load and the chain jumping sprocket teeth. In either case the failure is difficult to distinguish from a fatigue failure.

Rollers will very rarely fail in ultimate strength. Due to their position, the overload would be transferred to another component in the chain joint causing it to fail.

COMMON CHAIN PROBLEMS

The following is a discussion of the most common chain problems and a possible correction for each.

Chain Misalignment

Chain misalignment can cause a variety of problems. It can cause the drive to be noisy and have stiff chain joints, and causes wear on the sides of the sprocket teeth and insides of the chain link plates (Fig. 8-5). One of the best methods to check the

Figure 8-5. Wear on inside of roller link plates.

alignment is to use a piece of string as illustrated in Fig. 8-6. If the string doesn't touch on all four points indicated, adjustment should be made so that it does to insure proper alignment.

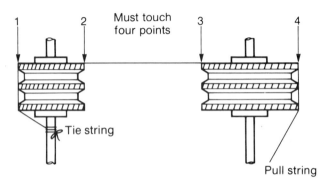

Figure 8-6. String alignment method.

Worn Out Chains

Worn out chains cause the drive to be noisy, but they'll also ride high on the sprocket teeth, even causing the chain to jump teeth in the drive. This may also cause broken chain components. The only solution is to replace the chain.

Improper Tension

Improper tension can cause the drive to be noisy and the chain to climb the sprocket teeth, or cause chain whip. Adjusting the tension to a 2% deflection of the center distance on the slack side will remedy the situation. (See Fig. 8-7.)

Improper Lubrication

Improper lubrication causes the chain to run hot, develop stiff joints, or run noisily. If a too thick lubricant is used, it may cause the chain joint to try to stay on the sprocket. This can be remedied by using the lubricant recommended by the manufacturer for your application. If the information cannot be found, a good rule of thumb is to use a 30 weight oil in a standard application, with an ambient temperature of 60–90°F. When operating outside this range adjust the grade of oil accordingly.

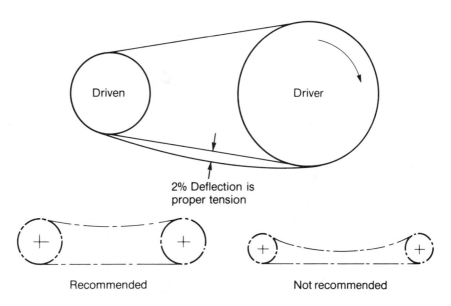

2% Deflection is
proper tension

Recommended   Not recommended

Figure 8-7. Slack side deflection.

ROLLER CHAIN SPROCKETS

Roller chain sprockets come in three basic designs: detachable hub, single hub, and double hub (Fig. 8-8).

Detachable Hub

The detachable hub uses some form of an incline plane principle to wedge the sprocket to the shaft. There's some form of allen screws or hex headed bolt to tighten the sprocket to the hub. Most manufacturers supply auxiliary tapped holes for the purpose of forcing the sprocket back off the hub. The various manufacturers usually provide a comprehensive selection of bushing sizes for the sprocket. This has the advantage of allowing the sprocket to be used on various sized shafting. This design is very practical where frequent installation and removal must be made.

Single and Double Hub

The single and double hub are used on a shaft with a key and set screws. They're usually bored to size, which limits their use

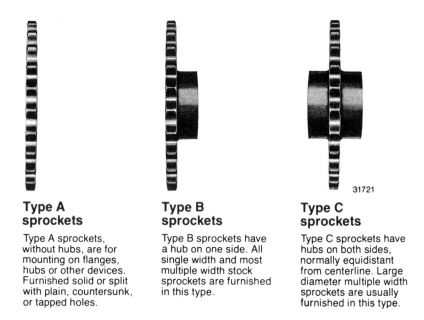

31721

Type A sprockets

Type A sprockets, without hubs, are for mounting on flanges, hubs or other devices. Furnished solid or split with plain, countersunk, or tapped holes.

Type B sprockets

Type B sprockets have a hub on one side. All single width and most multiple width stock sprockets are furnished in this type.

Type C sprockets

Type C sprockets have hubs on both sides, normally equidistant from centerline. Large diameter multiple width sprockets are usually furnished in this type.

Figure 8-8. Sprocket types. (Courtesy of P. T. Components Inc.)

to one shaft size only. The size of the transmitted load determines the choice of the single or double sprocket.

While roller chain is commonly used in industry, it must be maintained to make the investment in this form of power transmission. The repairman must give consideration to the type of drive, maintenance requirements, and the repair frequency. When these points are kept in mind the chain drive will deliver the long and trusted service life that it was designed to deliver.

SILENT CHAIN

Silent chain is another form of chain that is popular in power transmission (Fig. 8-9). It has the following advantages over roller chain:

- Higher operating speeds
- More efficient
- Longer life
- Quieter and smoother operation.

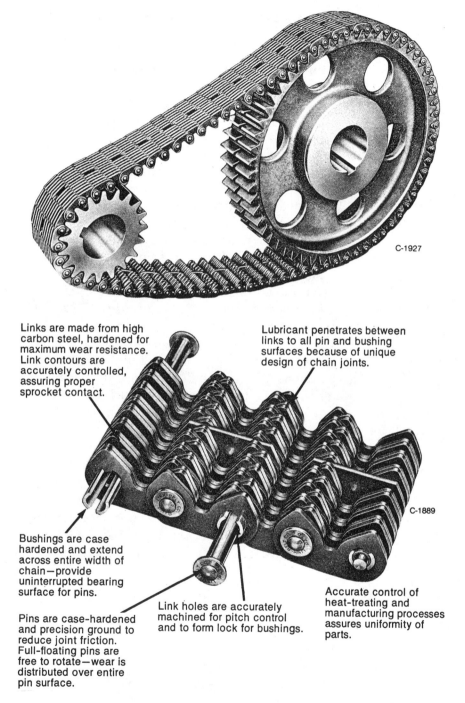

Links are made from high carbon steel, hardened for maximum wear resistance. Link contours are accurately controlled, assuring proper sprocket contact.

Lubricant penetrates between links to all pin and bushing surfaces because of unique design of chain joints.

C-1927

C-1889

Bushings are case hardened and extend across entire width of chain—provide uninterrupted bearing surface for pins.

Pins are case-hardened and precision ground to reduce joint friction. Full-floating pins are free to rotate—wear is distributed over entire pin surface.

Link holes are accurately machined for pitch control and to form lock for bushings.

Accurate control of heat-treating and manufacturing processes assures uniformity of parts.

Figure 8-9. Silent chain. (Courtesy of P. T. Components Inc.)

The drawback to silent chain is that it's more expensive than roller chain.

Unlike standard roller chain, silent chain has two dimensions that must be specified when ordering. In addition to the pitch, the thickness must be specified. For example:

SC-8-08
SC-silent chain
8-pitch in eighths $\frac{8}{8}$ or 1 inch
08-thickness in fourths $\frac{8}{4}$ or two inches thick.

In any silent chain installation, careful inspection should be given to the silent chain sprocket. If the silent chain sprocket is center guide or double center guide, then that's the only chain that can be used with it. When the thicknesses are the same then side guide chains can be interchanged.

Maintenance of silent chain is very similar to roller chain. Lubrication and alignment are of primary importance. In alignment, care should be taken to inspect chain and sprocket guide. Any tendency to wear on one side of the guide and sprocket will indicate an alignment problem, and prompt attention should be given to the drive. Lack of the proper lubricant will shorten the life of the drive to a fraction of what it's designed to deliver.

SUMMARY

Roller chain is a positive means of transmitting power between shafts with long center distances. The advantage they have over V-belts is positive drive. The advantage they have over gears is the flexibility. If the maintenance guidelines outlined in the chapter are followed, the chain drive will have a long service life with a minimum of maintenance problems.

REVIEW QUESTIONS

1. What are the two links that make up a standard roller chain?

2. What are the other two links that may be found in a roller chain drive, and what purpose does each serve?

3. Describe each of the following roller chain designations:
   (a) 35
   (b) 25
   (c) 60-H
   (d) 41
   (e) 100-3

4. List the seven steps to a roller chain installation.

5. What is the maximum elongation that a roller chain can tolerate before it should be replaced?

6. How many replacement chains will a sprocket usually outlast?

7. What causes roller fatigue?

8. What problems will chain misalignment cause?

9. What are the three basic hub designs for roller chain sprockets?

10. What advantage does silent chain have over roller chain?

11. What are the two most important factors in the life of a silent chain drive?

Chapter 9

Gears

In a belt drive the pulleys are driven by the frictional contact from the belt. Now picture the pulleys pressed together. The pulleys are not the ideal frictional surface and would have a problem with slippage. The slippage would be eliminated if the outer circumference could be cut or notched so that the pulleys would engage as they rotate. This is the basic idea behind a gear. The tooth design may vary from gear to gear; the principles are the same, however. Gears may be defined as two or more circular discs that transmit power or torque by the engaging of consecutive teeth.

Gear drives are among the earliest forms of power transmission devices. They're used to transmit power, change speed, or change rotation of shafts. They can be used to transmit power between shafts at right angles, between parallel shafts, or between shafts whose center lines don't intersect or aren't in the same plane. When gears have different numbers of teeth in mesh, the one with the smaller number of teeth is called the *pinion* and the one with the larger number of teeth is called the *gear*.

Gear drives are usually divided into two main classifications: those with parallel shaft axis and those with shafts that intersect at various angles.

GEAR DRIVES WITH PARALLEL SHAFT AXIS

The gear drives that have parallel shaft axes are:

- Spur
- Helical
- Herringbone
- Internal gear
- Rack and pinion

Spur Gear

The spur gear (Fig. 9-1) is a gear with the teeth cut parallel to the shaft axis. The spur gear is the basic gear and all other parallel shaft gears are derived from the spur. When the teeth are in contact or mesh the power is transmitted by a combined sliding and rolling action between the teeth. Fig. 9-2 illustrates the most common terms associated with the spur gear. The ease of manufacture, low cost, and ease of maintenance make the spur gear one of the most popular designs.

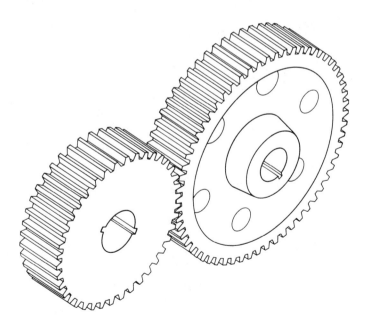

Figure 9-1. Spur gear.

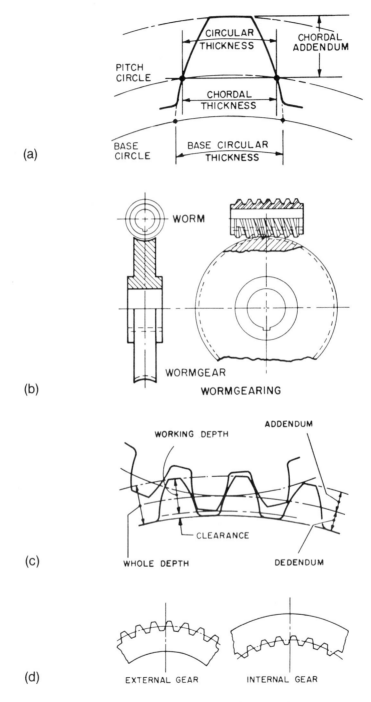

Figure 9-2. Basic gear terms.

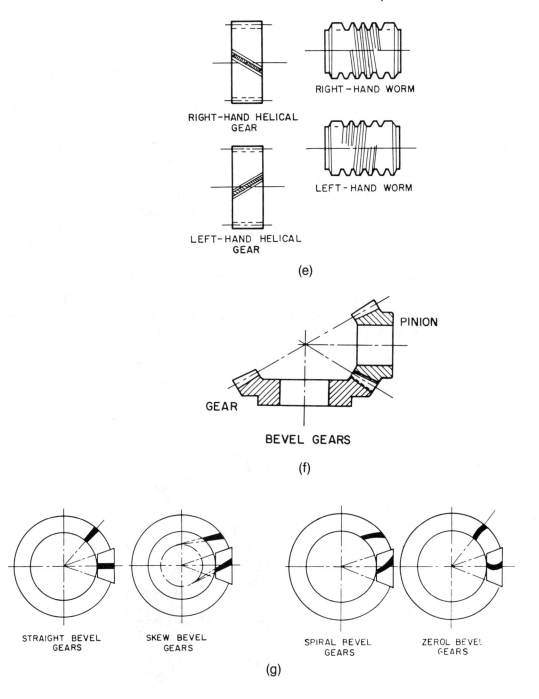

RIGHT-HAND HELICAL
GEAR

RIGHT-HAND WORM

LEFT-HAND WORM

LEFT-HAND HELICAL
GEAR

(e)

PINION

GEAR

BEVEL GEARS

(f)

STRAIGHT BEVEL
GEARS

SKEW BEVEL
GEARS

SPIRAL BEVEL
GEARS

ZEROL BEVEL
GEARS

(g)

Figure 9-2. Continued.

Helical Gear

The helical gear (Fig. 9-3) was designed from the spur to allow for higher speeds and higher loads. Instead of the teeth being cut parallel to the shaft axis, they're cut at an angle. The angle they're cut at is the helix angle. This creates more of a problem for there are many different helix angles, 7–23 degrees being the most common. For the gears to properly mesh, they must not only have the same helix angle, but must also be of different "hands." This means that one must have the teeth cut in the right hand direction and the other in the left hand direction.

Since the teeth are cut at an angle, there'll always be more tooth surface in contact when in mesh. This allows for more strength, and also allows for a quieter and smoother drive. This also creates a problem. The angle at which the teeth are cut produces an end thrust when the gears are rotated. The greater the helix angle, the greater the end thrust produced. Any application designed for helical gears must also have bearings and other drive components capable of withstanding the end thrust produced.

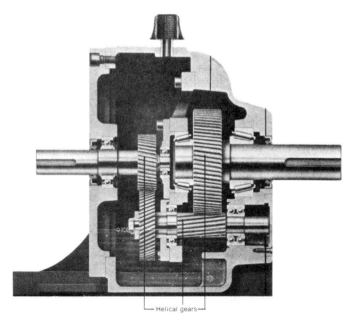

Helical gears

Figure 9-3. Helical gears. (Courtesy of P. T. Components Inc.)

One type of helical gear that doesn't fall into the parallel shaft gear is the crossed axis or the spiral gear (Fig. 9-4). These helical gears have a 45-degree helix angle and can be run on right angle drives. The difference in these drives is that the hand or the cut of the teeth must be the same to run at right angles. Since the teeth run at right angles, there isn't much surface contact of the teeth. This limits the use of this type of gear to only light loads.

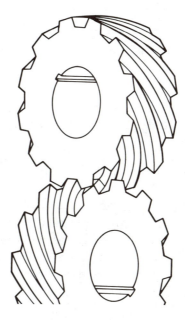

Figure 9-4. Spiral gears. (Courtesy of American Gear Manufacturers Association)

Herringbone Gear

The herringbone gear (Fig. 9-5) is basically a helical gear with both a right hand and a left hand set of teeth on the same gear. In fact, the herringbone gear is also called a double helical gear. These gear teeth have the advantage of having more tooth surface in contact, and so are stronger and quieter, and produce no end thrust. Because it has both left and right handed teeth, the two end thrusts that are generated cancel each other. The herringbone gear is the strongest and smoothest gear and is capable of higher speeds than any other parallel shaft gear. It's

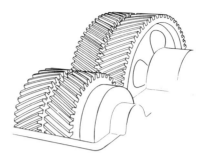

Figure 9-5. Herringbone gear. (Courtesy of American Gear Manufacturers Association)

also the most expensive gear to manufacture. For that reason you'll find it in application only where the spur or helical gears cannot be used.

Internal Gear

The internal gear (Fig. 9-6) has the gear teeth cut inside the gear, and a mating pinion that meshes inside the gear. The large outer gear is sometimes referred to as the *annular* gear. The gear drive may also have more than one internal gear. These types of drives are called *planetary* drives (Fig. 9-7). The small drive is called the *sun* gear, the others are called the *planets* and the *ring* (or annulus) gear. The teeth are usually of the spur cut, but may also be helical or herringbone.

Rack and Pinion Gear

The rack and pinion (Fig. 9-8) is used to convert rotary motion into linear motion or linear motion into rotary motion. This is accomplished by a flat rack with teeth cut into the face of the rack meshed with a pinion. The rack and pinion has either spur, helical, or herringbone teeth. The design requirements versus the cost determines what type will be used.

GEARS WITH RIGHT ANGLE APPLICATIONS

Right angle applications are divided into two basic categories, worm and bevel.

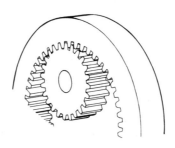

Figure 9-6. Internal gear. (Courtesy of American Gear Manufacturers Association)

Figure 9-7. Planetary gears.

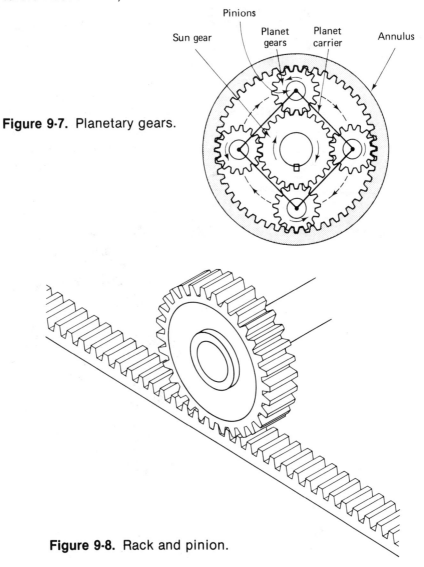

Figure 9-8. Rack and pinion.

Worm Gears

The worm gear (Fig. 9-9) is capable of transmitting large amounts of power. As the gear ratio increases, the efficiency decreases, which limits the use of the worm gear. In a worm drive there are two parts, the *worm* and the *worm gear*. The worm is usually made of steel or some stronger alloy, while the worm gear is usually made from brass or bronze or a softer alloy. The worm is usually the driver; however, drives having a ratio of less than 30:1 can be back driven. In applying gear ratios to worm drives, the ratio 30:1 means that for every 30 revolutions of the worm, the gear would turn once. Any worm gear drive with a ratio of 30:1 or greater is called a self-locking drive, meaning it cannot be back driven. To change the ratio in a drive, a worm gear with a different number of teeth can be installed. The worm can also affect the drive by having more than one start or lead on it. If it has more than one start (usually six is maximum) it turns the worm gear faster. For example, a 3-start or lead turns the gear 3 times as fast as a single.

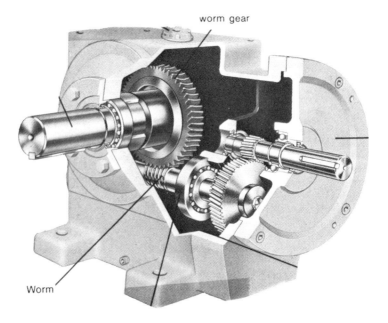

Figure 9-9. Worm and worm gear. (Courtesy of P. T. Components Inc.)

Bevel Gears

Bevel gears are very versatile as angle drives. There are four types:

- Straight
- Spiral
- Zerol
- Hypoid

Bevel gears (Fig. 9-10) have their teeth cut on cones. Depending on the type of gear, the teeth may be straight, spiraled, or a combination. The axis may intersect at a given angle or not at all. They're efficient and work where worm drives cannot.

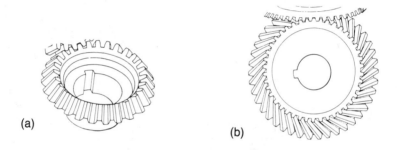

(a)

(b)

Figure 9-10. Bevel gears: (a) straight and (b) spiral. (Courtesy of American Gear Manufacturers Association)

Straight bevel gears have teeth that, if they were extended, would converge at the center of the gear. They usually run at shaft intersection angles of 90 degrees, but may be found at angles less than 90 degrees (acute) or more than 90 degrees (obtuse). If two straight bevel gears are in mesh and have the same number of teeth, and are at a 90 degree shaft intersection angle, they're called *miter gears*.

Spiral bevel gears have the same shape as straight bevel gears except the teeth have a helix angle. This type affords the same advantages as the helical gears. The spiral teeth allow for more tooth surface to be in contact at any time, and make a smoother and quieter drive.

Zerol bevel gears are a combination of the straight and the spiral bevel gears. They have teeth that are cut so they would intersect in the center of the gear, but they're cut on a machine that cuts the spiral teeth, so they're a straight spiral tooth. The advantage is that there is no thrust set up by the helix angle, but the shape of the tooth still provides more tooth surface in contact than in the straight bevel gear.

Hypoid gears (Fig. 9-11) are spiral bevel gears that have axes that don't intersect. This is a great advantage on heavier drives, for the shafts can be extended and support can be given each end of the shaft.

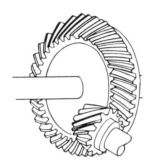

Figure 9-11. Hypoid gears. (Courtesy of American Gear Manufacturers Association)

LUBRICATION OF GEARS

In all gear drives, lubrication is a prime consideration. The function of the lubricant is to build a wedge to prevent actual metal-to-metal contact. Selection of the correct lubricant becomes very important. If the lubricant is too thin it is not able to separate the mating surfaces under load. The film barrier will rupture and metal-to-metal contact occurs. The surfaces in direct contact under load will generate intense heat. The surface asperities will weld together and then tear apart when coming out of mesh. This welding and tearing results in rapid deterioration of the gear tooth profile. If the lubricant is too thick it won't enter the meshing gears, and won't build the wedge with the same results.

Equally important is the proper filtration of the lubricating fluid. If foreign material is allowed between the teeth while in mesh, the material will be imbedded into the tooth surface, or it'll scratch the tooth surface as it slides into and out of mesh. If possible, a filter with a 3-micron rating is suggested for all recirculating lubrication systems.

Lubrication Methods

Lubrication systems for gear systems are divided into four categories: splash, slinger, drip, and spray.

Splash method. The splash method has a fluid level in an enclosed case high enough to be picked up by the lower gear. As the gear rotates it splashes oil in the case in sufficient quantity to lubricate the gears.

Slinger method. The slinger method has an eccentric disc located on the shaft. This disc dips into the oil, picking up sufficient quantity to lubricate the gears.

Drip and spray methods. The drip and spray methods are both used on higher speed drives. The system must have an external pump to provide flow of lubricant. The drip uses a small orifice to allow very slow feed of the lubricant to the desired location in the gearcase. The spray uses a higher flow rate allowing the maximum lubrication to the desired areas in the gearcase.

Backlash

Working hand-in-hand with the lubricant is the backlash in a gear drive. Backlash is the amount of clearance that the manufacturer designs into a gear. This clearance is designed into the gear itself (Fig. 9-12). It's just a small amount of material that is removed from the tooth surface. This enables the teeth to enter and leave the mesh without binding. Backlash also leaves clearance for the lubricant to enter the mesh to help protect the gear tooth.

GEAR TOOTH FAILURE

It's important to be able to recognize wear patterns on the teeth. This enables an individual to recognize potential problems before they become so severe that the drive is damaged or destroyed. It may also prevent the installation of replacement

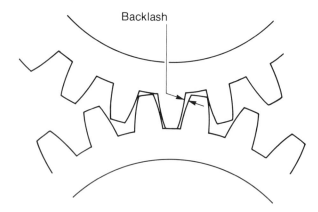

Figure 9-12. Backlash.

gears when a problem still exists. The American Gear Manufacturer's Association divides gear tooth wear and failures into four categories:

- Wear
- Surface Fatigue
- Plastic Flow
- Breakage

Wear

Wear is merely a normal function of the gear design where the metal is removed from the gear tooth in the contact areas of the tooth. There are eight basic types of wear (Fig. 9-13):

Normal. Normal wear results in a smoothing or polishing of the tooth surface. This wear ceases once the initial surface asperities are worn away. See Fig. 9-13 (a).

Moderate. This wear takes a longer time to develop. It results in removal of material from the addedum and dedendem, but leaves the pitch line virtually untouched. A higher viscosity lubricant usually stops this type of wear. If that doesn't work, increase in speed will help to build the wedge of lubricant. See Fig. 9-13 (b).

Destructive. This is a heavier wear that removes enough material to change the shape of the gear tooth. It's usually noticed by excessive noise during operation. This wear can be corrected by the same methods as for moderate wear. See Fig. 9-13 (c).

Abrasive.   This is identified by scratches in the tooth in the direction of the mesh. It's caused by particles in the lube. The prevention is to increase filtration. See Fig. 9-13 (d).

Scratching.   This is severe abrasion. It's caused by larger particles passing through the mesh and can be eliminated by better filtration or more frequent oil changes. (See Fig. 9-13 (e).

Scoring.   This is metal removal by welding and tearing. It can look like frosting or can have obvious weld and tear markings in the direction of the sliding mesh. It can be across the whole tooth or may be in a localized area of the tooth. Scoring is caused by failure of the fluid barrier to prevent metal-to-metal contact. A high viscosity lubricant or one with an extreme pressure additive is recommended. See Fig. 9-13 (f).

Interference.   This condition occurs where there is backlash in a set of gears. There's heavy wear at the root of the tooth. It can only be corrected by a design change. See Fig. 9-13 (g).

Surface Fatigue

Surface fatigue occurs when the tooth surface is stressed beyond its design limits. This results in the contacting asperities being fatigued and removed and leaves pits. This may continue until the tooth is destroyed. There are three types of surface fatigue (Fig. 9-14):

Initial pitting.   This type leaves very small pits, and is usually caused by some design flaw in the tooth. It generally occurs when the tooth profiles don't match correctly. It's usually self-correcting. See Fig. 9-14 (a).

Destructive pitting.   This generally results from a tooth running under loads higher than it is rated to carry. Reducing the load is the only cure. See Fig. 9-14 (b).

Spalling.   This is similar to pitting, except that the craters are very shallow. The spalling may advance so that the craters join, leaving larger ones. Spalling may continue until the tooth is destroyed. Reducing the load is the cure for spalling. See Fig. 9-14 (c).

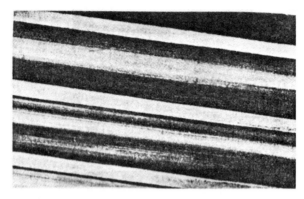

(a) Normal gear wear.

(b) Moderate gear wear.

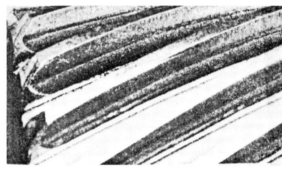

(c) Destructive gear wear.

(d) Abrasive gear wear.

Figure 9-13. Types of gear wear. (Courtesy of American Gear Manufacturers Association)

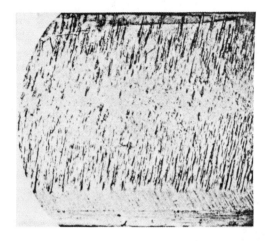

(e) Scratching.

(f) Scoring.

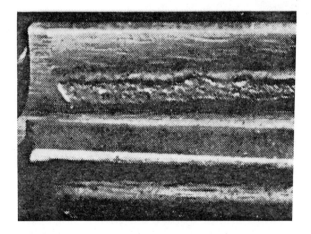

(g) Interference wear.

Figure 9-13 Continued.

(a) Initial pitting.

(b) Destructive pitting.

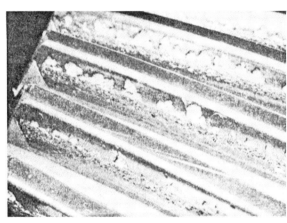

(c) Spalling.

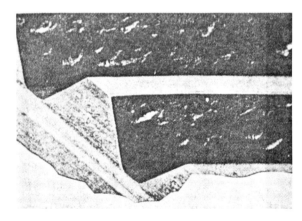

Figure 9-14. Surface fatigue. (Courtesy of American Gear Manufacturers Association)

Plastic Flow

Plastic flow is the movement of subsurface material on a gear tooth face. It's caused by high contact stresses or overloads. There are three basic types of plastic flow (Fig. 9-15):

Rolling and peening. This is a movement of material to the ends of the gear teeth. It's usually apparent by the fins at tips of the teeth. The rolling occurs with the sliding continuous overload. The peening occurs with heavier, hammering loads. This usually indicates the the drive is running with a too high load. See Fig. 9-15 (a).

Rippling. The gear has a fish scale appearance, caused by the movement of the subsurface material. This problem is again related to heavy loads, but may be remedied by a lubricant with an extreme pressure additive. See Fig. 9-15 (b).

Ridging. This also is caused by heavy loads. The condition appears as ridges in the gear tooth. The ridges run in the direction of the sliding of the mesh. It may even have the appearence of scratching. The best remedy is to reduce the load, or use a lubricant with an extreme pressure additive. See Fig. 9-15 (c).

Breakage

Breakage is the removal of a tooth or part of a tooth. It is caused by an overload of the gear drive—either one shock load or continuous overloading of the teeth. There are two basic types of breakage (Fig. 9-16):

Fatigue. This is caused by repeated overloads, resulting in small stress cracks, which eventually cause the tooth to break off the gear. Reducing the loads or redesigning the drive is the only cure. See Fig. 9-16 (a).

Overload. Overload results when one sudden shock load is severe enough to remove the tooth from the gear. This may be caused by foreign material wedging in the mesh causing an overload. Continuous monitoring of the loads and correcting the overloads is the only cure. See Fig. 9-16 (b).

By watching the types of gears and by carefully maintaining the gears, the life of the gears can be greatly extended. If failures

(a) Rolling and peening.

(b) Rippling.

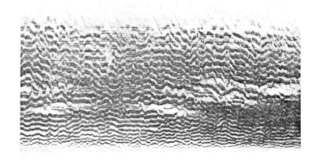

(c) Ridging.

Figure 9-15. Plastic flow. (Courtesy of American Gear Manufacturers Association)

(a) Fatigue.

(b) Overload.

Figure 9-16. Breakage. (Courtesy of American Gear Manufacturers Association)

do occur, the cause should be identified and corrected before a replacement is installed.

SUMMARY

Properly maintained gear drives will deliver almost infinite service lives. They're capable of changing speeds, changing direction, and changing torque in any drive. Maintaining a gear drive must involve careful inspections to observe wear patterns on the different gears. Knowing the wear patterns will enable a repairman to spot trouble before failure of the drive results.

REVIEW QUESTIONS

1. Into what two main classifications are gears usually divided?

2. What are five examples of parallel shaft drives?

3. What advantages does helical gearing have over spur gearing?

4. What advantage does herringbone gearing have over helical gearing?

5. What type of gear drive produces linear motion?

6. What are the two main divisions of right angle drives?

7. What is a self-locking worm gear?

8. How does the number of leads on a worm drive affect the speed of the drive?

9. What are the four main types of bevel gears?

10. What is backlash?

11. Why is lubrication important in a gear drive?

12. List the four divisions of gear tooth wear and an example of each.

Chapter 10

Couplings

Couplings are used to connect mechanical drives to a prime mover and are found in all industries. They fall into two broad classifications, mechanical and fluid. In this text, we'll discuss mechanical couplings.

Mechanical couplings can be further divided into three classifications, rigid, flexible, and universal joint (Fig. 10-1).

RIGID COUPLINGS

Rigid couplings are used where two shafts must be directly coupled. The alignment is critical, because there's no play in a rigid coupling. The two main types of rigid couplings are the flanged and the sleeve. *Flange* couplings are used in heavy duty applications, with large shafts and heavy loads. *Sleeve* couplings are used in smaller applications where shaft sizes are less than six inches.

FLEXIBLE COUPLINGS

Flexible couplings are made to compensate for some shaft misalignment. There are four types of misalignment that may be encountered in aligning a coupling.

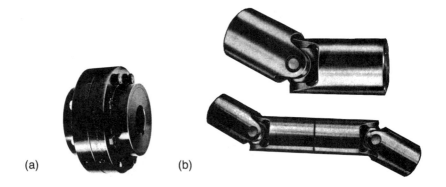

(a)   (b)

Figure 10-1. (a) Flange coupling, (b) single and double universal joints. (Courtesy of P. T. Components, Inc.)

- angular
- parallel
- combination angular parallel
- shaft float

Flexible couplings can compensate for one or, in some cases, all four types of misalignment. But flexible couplings are not cure-alls. Just because they say flexible, it doesn't mean they can compensate for ridiculous amounts of misalignment. If a manufacturer says its coupling can compensate for .005 misalignment, it still should be aligned as close as possible before start-up. The illustration might be made using a car speedometer. Just because the speedometer goes to 120 mph, you don't drive it that fast everywhere you travel. The same can be said for a coupling; just because it says .005, it doesn't mean it should run with that much misalignment.

Some of the more common flexible couplings (see Fig. 10-2) are the following. They all use play in the coupling to compensate for misalignment:

- gear
- grid
- chain
- spring
- paraflex
- jaw and spider

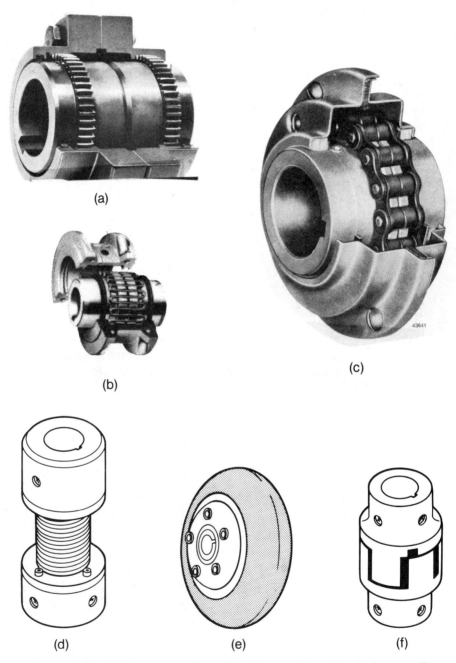

Figure 10-2. (a) Gear coupling, (b) grid coupling, (c) chain coupling, (d) spring coupling, (e) paraflex coupling, (f) jaw and spider coupling. (a, b, and c courtesy of P. T. Components, Inc.)

UNIVERSAL JOINTS

The universal joint is used where severe misalignment is encountered. There are two basic types of universal joints, the single and the double. The *single universal joint* can compensate for angular misalignment and the *double universal joint* can compensate for offset and parallel misalignment. If there's to be a change in the angle of misalignment, then a special slipped shaft, splined universal joint must be used.

INSTALLING COUPLING

When installing any coupling there are a variety of problems that may be encountered. The following is intended to be a guide to installing a coupling.

The first step is to determine if all components are in place and ready to assemble. This would include determining if the hub of the coupling is a clearance fit or an interference fit. A *clearance* fit has some means of locking the coupling on the shaft once it's slid into place. An *interference* fit means that the coupling hub has to be heated before it's installed to provide the proper clearance to get it on the shaft.

Heating the Coupling

There are two main methods for heating a coupling before it's installed, the oil bath and the air oven.

Oil bath. The oil bath is a faster way to heat the hub, but you're limited to a maximum temperature of 350°F. When using an oil bath, care should be taken that the hub doesn't rest on the bottom of the tank in direct contact with the heating source.

Air oven. The air oven is able to heat the hub to a higher temperature, but it's not as fast. Care should be taken not to allow the coupling to be heated above 600°F because above this the coupling steel will react with the air. One advantage of the air oven over the oil bath is that the coupling can be handled with heat resistant gloves, for it will not be slippery as it would in the oil bath.

Key Size

Another problem area in installing couplings is the key. If the key is not the correct size it will result in premature coupling failure. If the key is loose in the keyway it will, over a period of time, work back and forth in the keyway enough to begin wearing the keyway and the key. When this type of wear progresses enough, the coupling half may actually be able to turn on the shaft. When this occurs the coupling is useless and will have to be replaced. The other problem is having a too tall key in the coupling. (See Fig. 10-3.) This prevents the hub from having a proper fit on the shaft, and could set up stresses that split the coupling hub. The keys should have a tight sliding fit in the keyway when it's correctly installed.

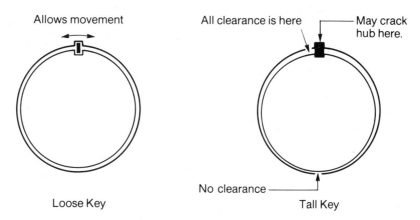

Figure 10-3. Loose and tall key.   *KEYWAYS STRAIGHT SLED RUNNER WOODRUFF*

Coupling Bolts

Coupling bolts are another problem area during installation. Coupling bolts should be torqued to the recommended values during installation. Too many times the mechanic tightens the bolts by feel, not applying enough torque to properly hold the bolt. Coupling manufacturers also provide special bolts with the couplings. These bolts are usually a grade 8 hardness. They should be replaced with a bolt of the same rating or a higher rating. If the bolts are replaced with a softer bolt, they'll stretch and fail prematurely. During installation of the coupling bolts you should turn the nuts and not the bolts. This will enable the correct amount of tension to be applied to the bolt. However, when removing a bolt from a coupling, it's best to loosen the bolt and not the nut. If a lockwasher is on the assembly and the nut is turned, it'll dig into the coupling and damage the surface the lockwasher is in contact with.

Lubrication

When installing the coupling, it's important to notice where the fittings for lubrication are located. If possible when assembling the coupling, position the fittings 180 degrees apart (Fig. 10-4). This'll make relubrication of the coupling easier. By removing both fittings for the lubrication, and positioning them horizontally, the lubricant can be added till it runs out the other side, indicating it is half full, the proper level for most flexible gear couplings.

Alignment of Coupling Halfs

Alignment of coupling halfs can be accomplished in a five-step procedure.

1. Vertical angular
2. Vertical parallel
3. Horizontal angular
4. Horizontal parallel
5. Recheck all steps

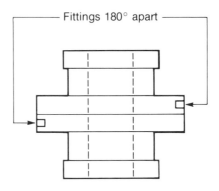

Figure 10-4. Position of grease fittings.

Vertical angular. The vertical angular (Fig. 10-5) can be checked by placing the dial indicator on the driving shaft. Place the tip of the indicator against the other coupling half. Lock the shafts together and rotate the shafts 180 degrees. Notice the difference in the reading on the indicator. This is the amount of misalignment in the vertical angular direction. To determine the size of shims required, use this formula:

$$\frac{\text{distance between the drivers' base bolts}}{\text{diameter of the coupling half}} \times \text{the indicator reading} = \frac{\text{the thickness of the shims required}}{}$$

Once this is known, place the shims under the base and tighten the base down, and check the reading again to insure that it was correct. Now you're ready to go to the vertical parallel procedure (Fig. 10-6).

This time position the dial indicator so that it reads the difference in height of the two coupling halves. This can be read by positioning the indicator on the rim of the coupling. Set at zero, and rotate it 180 degrees. Take half the difference and place that size shim under the base of the driver. Recheck the reading to be sure the shims were the correct size.

Horizontal angular. The horizontal angular (Fig. 10-7) is read by placing the dial indicator to read the inside of the coupling halves. This time no shims are required. The shifting of the driver sideways is all that's required to adjust the readings.

Horizontal parallel. The horizontal parallel (Fig. 10-8) is read on the outside of the coupling half. It's adjusted by shifting the driver until it's in alignment.

The fifth step is to go back and recheck the previous four steps. This may seem time consuming, but it pays off in extended service life for the coupling.

SUMMARY

Couplings are engineered to give good service life. However, most couplings are destroyed by poor installation and mainte-

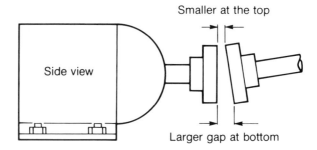

Figure 10-5. Vertical angular.

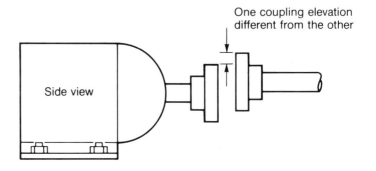

Figure 10-6. Vertical parallel.

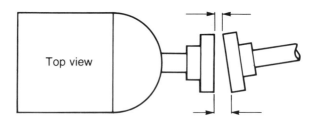

Figure 10-7. Horizontal angular.

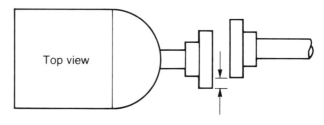

Figure 10-8. Horizontal parallel.

nance practices. If the procedures outlined in this chapter are followed, the coupling will deliver the service it was intended to deliver when it was manufactured.

REVIEW QUESTIONS

1. What are the two main classifications of couplings?

2. What are the three main divisions of mechanical couplings?

3. What type of mechanical coupling should be used when slight misalignment is encountered?

4. What type of mechanical coupling should be used where severe misalignment is encountered?

5. What two methods can be used to heat a coupling?

6. What two problems will an incorrect key cause in a coupling installation?

7. Why is using the correct bolts important in a coupling installation?

8. What is the five-step procedure for properly aligning couplings?

Chapter 11

Fluid Power Fundamentals

A discussion of industrial power transmission systems wouldn't be complete without including fluid power systems. Fluid power systems are used for simple fluid transfer to advanced drive systems. As with any subject it's always best to start with the very basics and build to the more advanced systems.

Every fluid power system must use a fluid medium. There are two very important terms in dealing with fluids, viscosity and density.

VISCOSITY

Viscosity is the fluid's resistance to flow. The higher the viscosity of a fluid the greater its resistance to flow. Conversely, the lower the viscosity the easier the fluid will flow. For example, cold molasses has a high viscosity and hot water has a low viscosity. With any fluid the colder it is, the higher its viscosity. This becomes important as the fluid power system undergoes temperature changes.

Measuring Viscosity

There are several methods of measuring a fluid's viscosity. The most common is the Saybolt Universal Seconds (SUS) (Fig. 11-1). In this test, a certain amount of fluid must pass through a certain sized orifice. The time it takes to pass through (measured by a stopwatch) is the fluid's SUS rating. The tests are usually conducted at temperatures of 100°F and 210°F. Viscosity plays a very important part in the pump's ability to move the fluid, which will be discussed later.

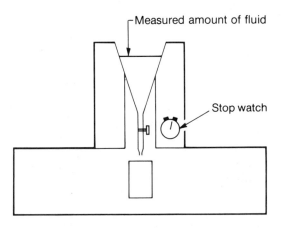

Measured amount of fluid

Stop watch

Figure 11-1. Saybolt universal seconds (SUS) test arrangement.

DENSITY

Density is a term that refers to the ratio of the fluid's weight to its volume. Density can be expressed in the following formula: $W = D \times V$, where

need to know hyd /pn... applications

W = weight
D = density
V = volume

Note: The units of volume and density must be consistent. For example, if density is in lbs/cu ft then volume must be in cu ft not cu in.

Density is a way to compare the heaviness of a given material

Table 11-1

		Density Pounds Per Cu. Ft.	Density Pounds Per Cu. In.	Specific Gravity
SOLIDS	Aluminum	166	.096	2.67
	Asbestos	153	.088	2.45
	Brass	532	.308	8.55
	Brick	125	.072	2.00
	Cement	90	.052	1.45
	Concrete	145	.084	2.43
	Copper	556	.322	8.93
	Cork	15	.008	0.24
	Glass	164	.094	2.60
	Iron (cast)	450	.260	7.21
	Iron (wrought)	485	.287	7.78
	Lead	710	.410	11.35
	Marble	170	.098	2.73
	Mercury	848	.491	13.60
	Steel	488	.283	7.83
	Stone (granite)	160	.093	2.78
	Timber (oak)	45	.026	0.72
	Timber (soft pine)	25	.014	0.40
LIQUIDS	Mineral Oil	55.6	.032	0.89
	Water-oil emulsion	56.2	.033	0.90
	Water-glycol solution	68.6	.040	1.10
	Phosphate ester	68.6	.040	1.10
	Silicone oil	64.8	.038	1.04
	Castor oil	60.5	.035	0.97
	Ethyl alcohol	49.4	.029	0.794
	Ethylene-gycol	69.9	.040	1.12
	Glycerol	78.6	.045	1.26
	Linseed oil	58.8	.034	0.944
	Mercury	849	.491	13.6
	Mineral oil, SAE 10	56.7	.033	0.91
	Olive oil	57.1	.033	0.915
	Turpentine	54.3	.031	0.872
	Water, fresh	62.35	.036	1.0
	Water, salt	64.0	.037	1.03

(Table 11-1). For example, cork has a density of 15 lbs/cu ft, and water has a density of 62.4 lbs/cu ft. If a material is less dense than another, it'll float or ride on top of the material with the higher density. This is the reason that a cork will float in water.

Specific Gravity

Another method of comparing one material with another is specific gravity. Specific gravity is the ratio of the density of a substance compared to the density of fresh water. The formula is:

$$\text{Sp Gr} = \frac{\text{Density of substance}}{\text{Density of fresh water}}$$

One point to keep in mind is that specific gravity applies only to liquids and solids. Gases change density when exposed to different pressures. Density and specific gravity can be determined for gases, but the pressure that the calculations were figured under must be stated.

PRESSURE

Pressure is a term that must be considered in fluid power. There are three ways that pressure can be created:

1. By the weight of a column of fluid. – STATIC
2. By the force applied to a column of fluid. – PRESSURE
3. By resistance to flow. - finge over hose

Weight of a Column of Fluid

The formula for pressure is:

$$\text{pressure} = \frac{\text{force}}{\text{area}}$$

Pressure is defined as force per unit of area. This can be further illustrated by examining a one-cubic-foot block of steel sitting on the floor (Fig. 11-2). The density of steel is 488 lbs/cu ft. Since our block is one cubic ft, it weighs 488 lbs. Since the block is resting on its side, the weight of the block is acting on the

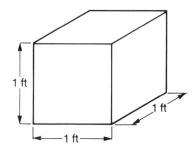

Figure 11-2. Weight on a given area.

floor on an area of 1 ft × 1 ft or 1 square foot. The pressure then becomes:

$$\text{pressure} = \frac{488 \text{ lb}}{1 \text{ sq ft}}$$

or if the conversion is made to square inches (Fig. 11-3), it becomes:

$$\text{pressure} = \frac{488 \text{ lbs}}{144 \text{ sq in.}}$$

or 3.39 lbs/sq in.

or 3.39 pounds/sq in.

or 3.39 psi

The unit psi is the most used unit in fluid power. It means pounds per square inch.

Force Applied to a Confined Fluid

All substances, including fluids, exert a force similar to the steel block. With fluids, it's possible to measure the pressure in any given point in the height of the column. The pressure exerted by a column of fluid can be determined if the height of the column and the specific gravity of the fluid are known. The formula is:

P(of the fluid) = .433 psi/ft × sp gr (of the fluid)

× height (of the column of fluid in feet)

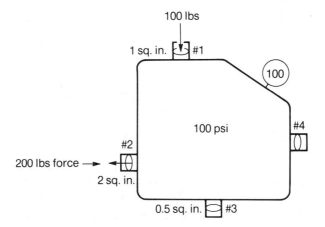

Figure 11-3. How force on a confined fluid creates pressure.

For example: determine the pressure at the bottom of a tank of mineral oil (sae 10) that is 25 feet high.

$$P = \frac{.433 \text{ psi}}{\text{ft}} \times .91 \times 25 \text{ ft}$$

$$P = 9.85 \text{ psi}$$

The concept of pressure is important to the understanding of how fluid power systems function. One concept of pressure is stated in Pascal's law: *Pressure on a confined fluid is transmitted undiminished and in all directions, and acts with equal force on equal areas and at right angles to them.* This principle is best illustrated by a closed container that has four plungers (Fig. 11-3). The force is applied to plunger #1 (a force of 100 lbs). Plunger #1 has an area of 1 sq in. Plunger #2 has an area of 2 sq in., and plunger #3 has an area of .5 sq in. Plunger #1 has applied to it a force of 200 lbs.

$$p = f/a, \text{ so } f = p \times a$$

for plunger #1.

$$p = 100/1 \text{ or } 100 \text{ psi}$$

The 100 psi is the pressure inside the closed container. The force on plunger #2 is:

$$f = 100 \text{ psi} \times 2 \text{ sq in., or}$$
$$f = 200 \text{ lbs.}$$

The force on plunger #3 is: $f = 100 \text{ psi} \times .5 \text{ sq in. or } f = 50 \text{ psi.}$

Resistance to Flow

Anytime flow is restricted, the pressure increases. The fluid always flows from a point of high pressure to a point of lower pressure. The fluid always flows through the path of least resistance. In fluid power circuits there are two types of flow paths, series and parallel (Fig. 11-4). In series flow paths the pressures add (Fig. 11-5). In parallel flow paths the flow takes the path of least resistance (Fig. 11-6). A device that's used to control flow or create pressure is called an orifice. The orifice may be a restricted passage in a pipe or a valve. The amount of flow through an orifice is determined by three things:

1. Size of the orifice
2. Viscosity of the fluid
3. Pressure drop

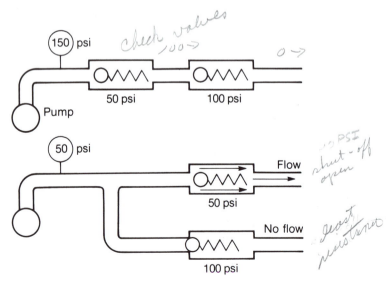

Figure 11-4. Series-parallel flow paths.

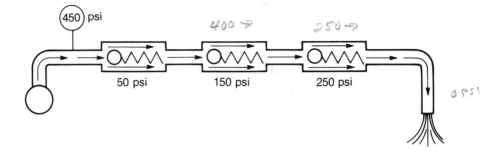

Figure 11-5. Series flow path

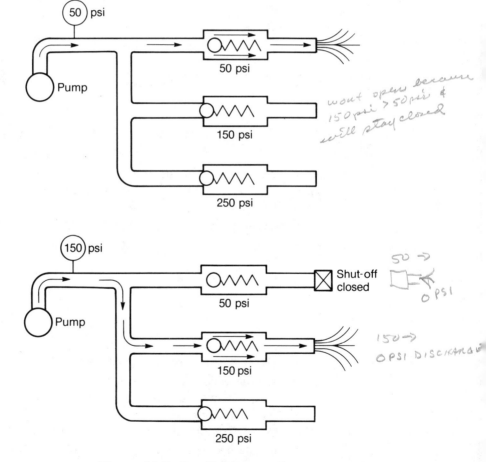

Figure 11-6. Parallel flow paths.

PRESSURE SCALES

One final point to consider in fluid power systems is the pressure scales. A column of fluid exerts a pressure at the bottom of the column. The atmosphere is a fluid and it exerts a pressure at the surface of the earth. At sea level that pressure is 14.7 psia (pounds per square inch absolute). The psia scale starts at a perfect vacuum (0 psia). Atmospheric pressure is equal to 14.7 on this scale.

Table 11-2

Atmospheric Pressure at Sea Level	PSIA Scale 14.7	PSIG Scale 0	Inches of Mercury 29.92	Feet of Water 34
2 Atmospheres	29.4	14.7	60 inches	74 feet
3 Atmospheres	44.1	29.4	90 inches	111 feet

GAUGE SCALES

Another scale is the gauge scale. Gauge scale ignores atmospheric pressure. Atmospheric pressure would be 0 psig. Table (11-2) gives the conversions for popular pressure scales. A timely conversion for the psia to psig scales is as follows:

psig + 14.7 = psia
psia − 14.7 = psig

PUMP INLET

Atmospheric pressure becomes very important when we consider the beginning of a fluid power system: the pump inlet. The fluid is not sucked into the inlet of a pump as is often thought. Atmospheric pressure acting on the surface of the fluid forces the fluid into the pump inlet, which is a place of lower pressure. This means that there'll be flow from an area of higher pressure (the reservoir) to an area of lower pressure (the pump inlet) (Fig. 11-7).

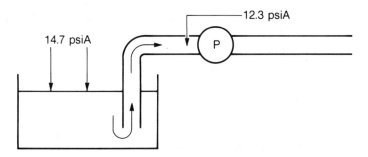

Figure 11-7. Flow from area of higher pressure to area of lower pressure.

Cavitation

If it were possible to pull a perfect vacuum at the inlet of a pump, you could lift a column of fluid equal to 14.7 psia. This is impossible, and even if it weren't, you wouldn't want to do so. There's dissolved air in all fluids, and at different pressures the dissolved air comes out of the fluid. At any pressure less than 12.2 psia, the dissolved air will bubble out of petroleum oil. When the air bubbles are carried to the outlet of the pump, they are exposed to pressures greater than atmospheric pressure. Under this increased pressure, they implode back into the fluid (Fig. 11-8). This implosion will actually be severe enough to remove metal from the pump, causing rapid deterioration of the pump. Unless the problem is corrected rapid failure of the pump can be expected. The condition just described is called cavita-

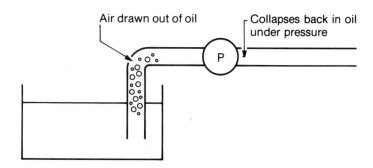

Figure 11-8. Cavitation.

tion. Cavitation is noticed by loud operating noises. In fact, it usually sounds as if the pump has gravel in it. Once the loud noise is observed, every effort should be taken to correct the condition before the pump is destroyed.

Inlet Air Leak

One condition that is often confused with cavitation is an inlet air leak. The inlet will allow the passage of some air from the atmosphere into the oil before it enters the pump (Fig. 11-9). Under the pressure on the outlet side of the pump, the air is dissolved into the oil, again resulting in the increased noise levels. The way to tell the difference between cavitation and an inlet air leak is to look in the reservoir. If it's cavitation, the oil will appear normal in color. If it's an inlet air leak, the air trapped under high pressure will bubble out in the reservoir. This leaves the oil with a foam on top or a milky color in the tank. Inlet air leaks are not as damaging as cavitation; however over a long period of time the results are the same. Both conditions should be corrected as soon as practical when they are detected.

Beginning fluid power systems is like coming to a "y" in a road. The systems branch off into two broad categories, hydraulics and pneumatics. Hydraulics will be covered in Chapter 12 and pneumatics will be covered in Chapter 13.

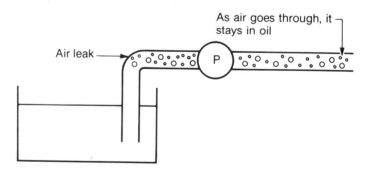

Figure 11-9. Inlet air leak.

REVIEW QUESTIONS

1. Explain why mineral oil floats on water.

2. What is the pressure at the bottom of a column of mercury 5 feet high?

3. What determines the amount of flow through an orifice?

4. What is the difference between absolute pressure and gauge pressure?

5. Explain the difference between cavitation and an inlet air leak.

6. Explain the difference between density and specific gravity.

Chapter 12

Basic Hydraulics

All hydraulic systems are comprised of basically the same components. The following material will present a broad overview of the components and how they function in the system.

RESERVOIR

All hydraulic systems begin with the reservoir (Fig. 12-1). The reservoir has several functions:

1. To store the fluid
2. To allow air to separate from the fluid
3. To allow contaminants to settle out of the fluid
4. To allow for temperature control of the fluid

The reservoir (Fig. 12-2) holds the fluid prior to its entry into the system through the pump. It's usually made of steel plates welded together. The tank usually has a drain at the bottom, to allow the fluid to be periodically changed. There's usually a removable endplate that can be used for access to clean the tank. The tank is also usually provided with an external sight glass to check the fluid level. The reservoir is equipped with a breather (unless the reservoir is pressurized) to allow the at-

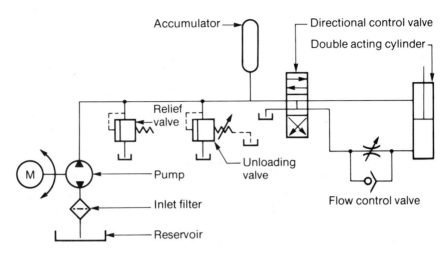

Figure 12-1. The reservoir of a hydraulic system.

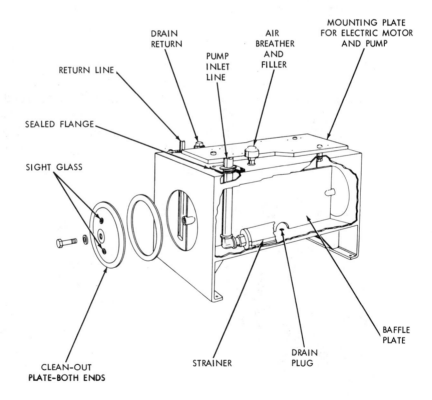

Figure 12-2. Typical hydraulic reservoir. (Courtesy of Sperry-Vickers)

mosphere to equalize the pressure inside the tank whether it's empty or full.

Inside the reservoir there's usually a baffle plate. The baffle plate prevents the returning oil from being drawn directly back to the inlet and sent back into the system. The baffle plate slows down the oil, and allows for any contaminants to settle to the bottom of the tank. This also helps to remove any entrapped air from the oil. By making the oil swirl around the tank, the oil is cooled before going back into the system.

Hydraulic systems that have a problem with heat are usually equipped with a heat exchanger (Fig. 12-3). This is usually a series of coils in the bottom of the reservoir that have cold water (or some other cooling medium) circulated through to remove the heat from the fluid. In some rare cases there are examples of heaters in the tank to warm the oil before it enters the system. This is usually found only in very cold climates where the viscosity of the oil needs to be lowered before it enters the pump.

FILTERS

The intake line to the pump may be equipped with a filter (Fig. 12-4) to prevent the pump from drawing contaminants into the system. This is one of the most common locations of the filter or strainer (a strainer is a coarse filter). Filters are rated in

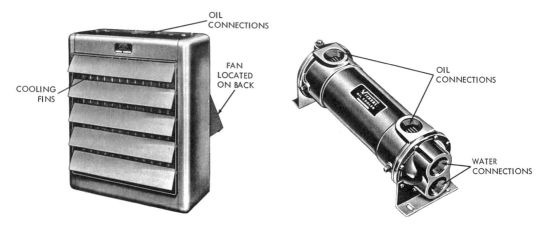

Figure 12-3. Heat exchangers (coolers). (Courtesy of Sperry-Vickers)

Figure 12-4. Typical inlet filter. (Courtesy of Sperry-Vickers)

microns (.000039 inch). The lower the micron rating of the filter, the smaller the particle it will remove. As the filter removes contaminants from the fluid, it begins to allow less and less fluid to pass through the filter element. Finally it won't allow enough flow for the pump to work properly. If it's not replaced before this time, the pump will begin to cavitate. This is an important reason for the filter to be serviced on a regular schedule.

Filters are rated on two systems, the nominal rating and the absolute rating. The *nominal rating* is the size of particle that the filter will stop most of the time. For example, a filter with a nominal rating of 30 microns will stop most particles 30 microns or larger. The *absolute rating* is the size of the smallest particle that can't pass throught the filter. For example, a filter with an absolute rating of 25 microns will stop all particles of the 25-micron size or larger (unless the contaminant particles are well rounded or long thin objects that may snake their way through the filter).

Other types of filters may be found in a hydraulic system, and they'll be considered as their location in the system comes along.

THE PUMP

The fluid passes through the filter and next enters the pump. Pumps can be classified as nonpositive displacement or positive displacement. *Nonpositive displacement* pumps are usually used in fluid transfer systems and aren't common in hydraulic systems. The nonpositive displacement pumps don't discharge a certain volume for each revolution of the pump. A *positive displacement* pump produces a given flow at a given speed. Positive displacement pumps are divided into three basic types (Fig. 12-5):

- vane
- gear
- piston

All three types of pumps work on the principle of expanding and decreasing volume.

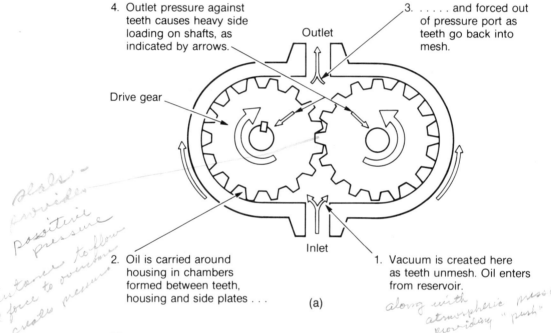

4. Outlet pressure against teeth causes heavy side loading on shafts, as indicated by arrows.

3. . . . . . and forced out of pressure port as teeth go back into mesh.

Outlet

Drive gear

Inlet

2. Oil is carried around housing in chambers formed between teeth, housing and side plates . . .

1. Vacuum is created here as teeth unmesh. Oil enters from reservoir.

(a)

plates — provides positive pressure

resistance to flow and force to overcome it creates pressure

along with atmospheric pressure providing "push"

Figure 12-5. (a) Typical gear pump, (b) typical vane pump, (c) typical piston pump. (Courtesy of Sperry-Vickers)

Vane Pumps

The vane pump consists of three basic parts, the *vanes*, the *rotor*, and the *housing*. As the rotor turns, the vanes extend out and contact the housing. The vanes may be spring loaded or may depend on the centrifugal force of the turning rotor to hold them out against the cam ring during start-up. After the pump

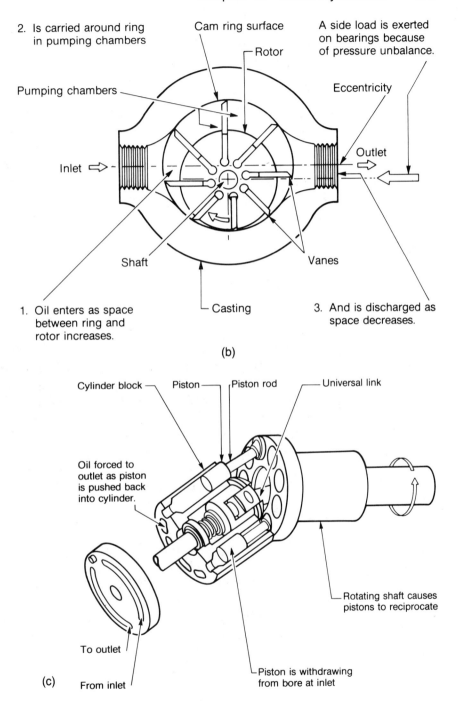

2. Is carried around ring in pumping chambers

Cam ring surface

Rotor

A side load is exerted on bearings because of pressure unbalance.

Pumping chambers

Eccentricity

Inlet ⇨

Outlet ⇨

Shaft

Vanes

1. Oil enters as space between ring and rotor increases.

Casting

3. And is discharged as space decreases.

(b)

Cylinder block

Piston

Piston rod

Universal link

Oil forced to outlet as piston is pushed back into cylinder.

Rotating shaft causes pistons to reciprocate

To outlet

From inlet

Piston is withdrawing from bore at inlet

(c)

Figure 12-5. Continued.

develops flow, the downstream pressure is used to hold the vanes against the cam ring. The chamber created by the vanes extending against the cam ring varies in size as the unit is rotated. The chamber increases in size as it approaches the inlet and decreases in size as it approaches the outlet. This decreasing chamber forces the fluid out of the chamber to the outlet.

Some vane pumps are adjustable volume. This style of pump allows you to move the rotor assembly to increase or decrease the chamber size. This, then, adjusts the amount of fluid that's forced out the discharge of the pump.

There are also *double vane* pumps. These pumps are actually two different pumps in the same housing. They're driven by the same shaft. The pumps usually come in different sizes, one stage used for low volume and the other used for high volume. This allows the designer some versatility in setting speeds of system components.

Gear Pumps

Gear pumps are two gears that are driven together. As the gears pass the inlet they pick up fluid that's trapped between the gear and the housing as the gear continues to rotate. The gear rotates to the outlet, where the trapped fluid is forced out of the outlet by the meshing of the two gears. The output of the gear pump can't be adjusted as can the vane. The only way to change the output is to speed up the prime mover driving the pump. The size of the chambers can't be varied so the output will be directly proportional to the speed the pump is driven.

Piston Pump

The piston pump consists of a series of cylinders moving in and out of chambers. The cylinders retract as they approach the inlet, allowing the fluid to fill the cylinder chamber. As they approach the outlet, the cylinder moves back in forcing the fluid out of the cylinder. There are several types of piston pumps (bent, axis, and radial) but the principles are the same. The volumetric output of the pumps can be varied by increasing or decreasing the stroke of the piston.

PUMP FLOW AND PRESSURE

One important point to remember about all pumps: pumps develop flow, not pressure. The pump displaces fluid, and if the outlet were left open would displace fluid under almost zero pressure. The restriction in the system or its resistance to flow is what develops the pressure. If the outlet of the pump is completely sealed, the pressure would theoretically increase till the pump destroyed itself. *fixed = constant flow*
variable = varied flow

Pressure Relief Valve

This problem is prevented in the system by a pressure relief valve (Fig. 12-6). This valve is usually installed immediately after the pump. Its purpose is to allow the system pressure to build to a certain level and after the pressure reaches that level, to dump the fluid to a tank. The valve is usually spring loaded, with a screw adjustment. As the screw is tightened, it compresses the spring, requiring higher pressure to open the valve. The problem with a relief valve is that when fluid is dumping to the tank through it, it is doing so under pressure, and the stored energy is converted to heat. This heat is transferred to the fluid and then must be dissipated in the tank. A solution to this problem is an accumulator (Fig. 12-7).

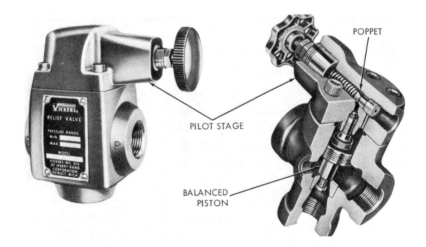

Figure 12-6. Pressure relief valve. (Courtesy of Sperry-Vickers)

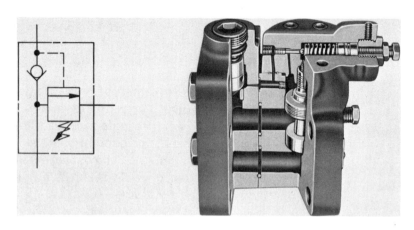

(a)

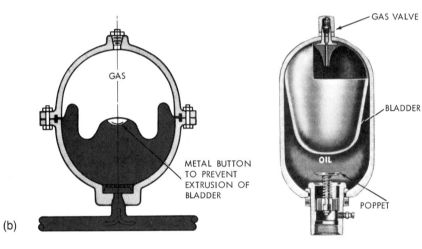

GAS

METAL BUTTON
TO PREVENT
EXTRUSION OF
BLADDER

GAS VALVE

BLADDER

OIL

POPPET

(b)

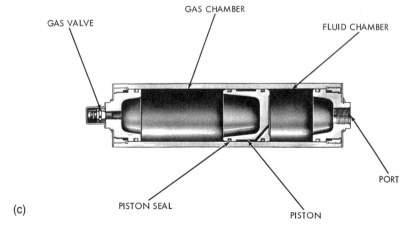

GAS VALVE

GAS CHAMBER

FLUID CHAMBER

PORT

PISTON SEAL

PISTON

(c)

Figure 12-7. (a) Unloading valve, (b) diaphragm accumulator, (c) piston accumulator. (Courtesy of Sperry-Vickers)

Accumulator and Unloading

The accumulator is to the hydraulic system as the battery is to an electrical system. The accumulator is a storage unit that holds a given volume of fluid under pressure. When the system requires flow, the fluid stored in the accumulator provides it. The pump runs the entire time but its flow is diverted to the tank across an unloading valve. When the system pressure lowers to a given point, the unloading valve closes and flow is provided to the system until a certain pressure is reached and the accumulator is recharged.

When these conditions are met, a pilot line to the unloading valve opens the valve, dumping the pump flow to the tank. The advantage of this over the relief valve is that it dumps under low pressure, which means less heat will be generated. This protects the system components from heat build-up that would be likely with just a relief valve.

Check Valve

One valve that is used in the diagram that was not explained is the check flow valve. The check valve (Fig. 12-8) allows flow in one direction and blocks flow in the other direction. It's easier to picture if you visualize the ball pushing into the v-shaped holder of the valve, blocking the flow. Then when the flow is reversed the ball is pushed off its seat allowing the flow to pass by the ball.

Directional Control Valve

The directional control valve (d.c.v.) (Fig. 12-9) is the next valve in the circuit. It's used to control the direction of the flow to the actuator. The most common directional control valves are the four-way directional control valves. They come in the 2-position or the 3-position valve. The two end envelopes in the d.c.v. are used to change the direction of the flow. The middle envelope is called the center.

There are four basic types of center conditions in a d.c.v. (Fig. 12-9) each with its own advantage. They are:

- open
- closed
- tandem
- float

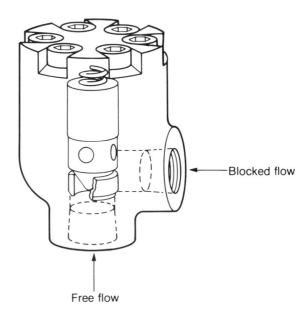

Blocked flow

Free flow

Figure 12-8. Right angle check valve. (Courtesy of Sperry-Vickers)

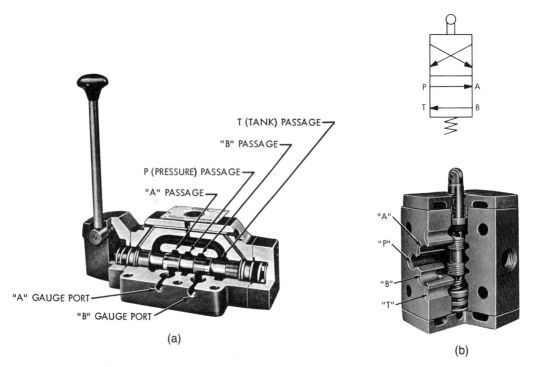

T (TANK) PASSAGE

"B" PASSAGE

P (PRESSURE) PASSAGE

"A" PASSAGE

"A" GAUGE PORT

"B" GAUGE PORT

(a)

P ——— A

T ——— B

"A"

"P"

"B"

"T"

(b)

Figure 12-9. Directional control valves: (a) manual 4-way d.c.v., (b) mechanically operated d.c.v. (Courtesy of Sperry-Vickers)

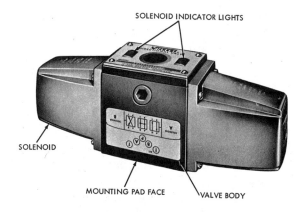

SOLENOID INDICATOR LIGHTS

SOLENOID

MOUNTING PAD FACE   VALVE BODY

Figure 12-10. Solenoid operated directional control valve. (Courtesy of Sperry-Vickers)

Open center.   The open center allows the flow to return to the tank under low pressure without an unloading valve. The disadvantage of the open center is that once it's centered the pump can't be used to power any other actuator. That's why it'll be used in a circuit that activates only one actuator.

Closed center.   The closed center blocks all passages in the valve. This center can be used in circuits with more than one actuator. The actuators may act independently of each other in this system. The disadvantage of this type of center is that the pump flow must be diverted by another means (relief valve, unloading valve). This valve also allows for some leakage around the spool, which may cause an actuator without a load on it to creep.

Tandem center.   Tandem center valves are used to unload the pump to the tank and at the same time block any flow to the actuator. The problem with this type of center is that the passages are so small, there's a significant pressure drop across the valve, which means generated heat. To compensate for this some manufacturers enlarge the lands on the spool for the pressure and tank ports. This decreases the size of the ports to the actuator, which restricts the flow and results in slower operation, but does keep the center cooler.

Float center.   The float center blocks the pressure port and vents the actuator lines to the tank. This allows the actuator to drift if necessary. The disadvantage to this type of valve is that the actuator cannot be stopped or held in any position. It does eliminate any pressure build-up across the lands on the spool.

Actuating the Directional Control Valve

There are several ways that a directional control valve can be actuated. One of the most common methods is with the *electric solenoid*. It uses an electrical current fed through a coil in the valve to set up a magnetic field to shift the spool (Fig. 12-10).

Another common way is with a *pilot operated* d.c.v. These valves use a pressure from another part of the system (or from an external source) to shift the spool (Fig. 12-11).

The other ways are *manual*, by using a foot pedal, a hand operated control, a cam operated switch, or some type of push button (Fig. 12-12). They all have their purpose in industry. Their placement in the system is usually chosen by the original equipment manufacturer.

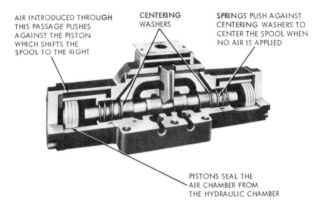

AIR INTRODUCED THROUGH THIS PASSAGE PUSHES AGAINST THE PISTON WHICH SHIFTS THE SPOOL TO THE RIGHT

CENTERING WASHERS

SPRINGS PUSH AGAINST CENTERING WASHERS TO CENTER THE SPOOL WHEN NO AIR IS APPLIED

PISTONS SEAL THE AIR CHAMBER FROM THE HYDRAULIC CHAMBER

Figure 12-11. Pilot-operated directional control valve. (Courtesy of Sperry-Vickers)

Flow Control Valves

The next valve in the line is the flow control valve (Fig. 12-13). This valve controls the flow, so it will affect the speed of the actuator. The valve is an orifice which may or may not be adjustable, with a check valve in parallel with it. The flow is controlled

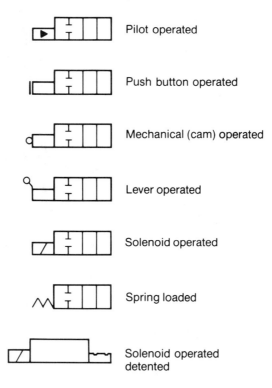

Pilot operated

Push button operated

Mechanical (cam) operated

Lever operated

Solenoid operated

Spring loaded

Solenoid operated detented

Figure 12-12. Symbols of directional control valve (DCV) actuators.

in one direction and is allowed to by-pass the valve in the other direction. This valve can be used in two ways in a circuit. It can either meter in or meter out. The *meter in* controls the flow into the actuator and allows free flow out of the actuator. The *meter out* allows full flow into the actuator, but restricts the flow out of the actuator.

Actuator

The final item in our imaginary hydraulic circuit is the actuator. The actuator can be either a cylinder or a motor.

Hydraulic cylinders. There are three types of hydraulic cylinders: single acting, double acting, and double rod (Fig. 12-14). The *single acting* applies force only in one direction. It must be retracted by another force, whether spring, gravity, or

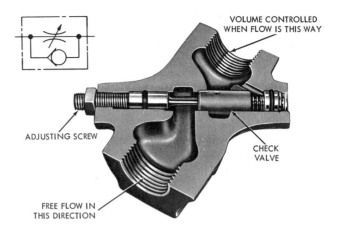

Figure 12-13. Flow control valve with by-pass. (Courtesy of Sperry-Vickers)

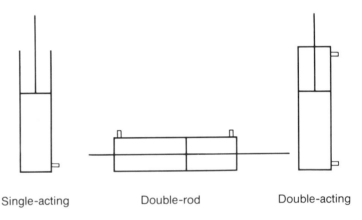

Single-acting   Double-rod   Double-acting

Figure 12-14. Three types of hydraulic cylinders.

the load. The *double acting* cylinder extends or retracts with force. The cylinder extends with more force than it will retract, because the rod end of the cylinder piston has less area for the pressure to act upon (because of the cylinder rod). The *double rod* cylinder has two rods so it extends and retracts with equal force.

Hydraulic motors. Hydraulic motors can be of two types, *reversible* or *nonreversible* (Fig. 12-15). The motors may also be classified by the driven element: vane, gear, and piston. The

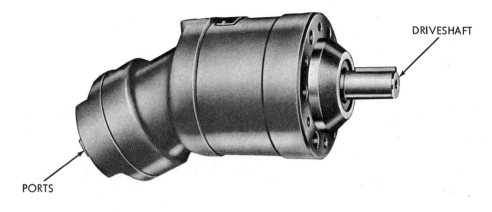

DRIVESHAFT

PORTS

(a)

5. As the piston passes the inlet, it begins to return into its bore because of the swash plate angle. Exhaust fluid is pushed into the outlet port.

4. The pistons, shoe plate and cylinder block rotate together. The drive shaft is splined to the cylinder block.

Piston sub-assembly

Swash plate

Outlet port

Inlet port

Drive shaft

Shoe retainer plate

1. Oil under pressure at inlet

2. Exerts a force on pistons, forcing them out of the cylinder block.

3. The piston thrust is transmitted to the angled swash plate causing rotation

(b)

Figure 12-15. (a) Bent axis piston motor, (b) in-line piston motor. (Continued on next page.)

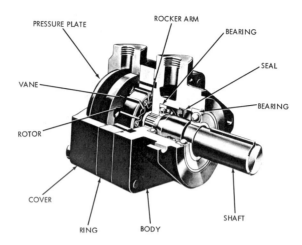

Figure 12-15. Continued. (c) Vane motor. (Courtesy of Sperry-Vickers)

motor works just the reverse of the pump. The fluid tries to get from an area of high pressure (inlet) to an area of low pressure (outlet). In doing so, the motor shaft is turned. The greater the flow coming into the motor the faster speed it will turn.

SUMMARY

In this brief overview of the basic hydraulic system, we have considered the primary system components. There are many other, newer styles of valves and devices that'll continually be changed and updated to improve the performance of the system. The increasing technology in hydraulics makes it a field that offers limitless potential for advancement, but along with that comes the challenge to learn.

REVIEW QUESTIONS

1. What purpose does a reservoir serve in a hydraulic system?

2. Why is there a baffle in a hydraulic reservoir?

3. What must the fluid pass through before it reaches the pump?

4. What are the three basic types of positive displacement pumps?

5. What purpose does a relief valve provide in a hydraulic system?

6. What purpose does an actuator serve in a hydraulic system?

7. What are the four basic types of center conditions in the directional control valves of a hydraulic system?

8. What purpose do flow control valves serve in the hydraulic system?

9. What are the two basic types of flow control arrangements in a hydraulic system?

10. What are the two basic types of actuators in a hydraulic system?

Chapter 13

Basic Pneumatics

Air has existed since the beginning of time. Man has harnessed the power of the wind to power many devices. Only in the last century has man used air as a power transmission medium. A brief comparison of a pneumatic system with a hydraulic system is shown below.

Pneumatic System

Uses air under pressure
Is a low power system
Transmitting medium is compressible
Leaks are clean
Is a low pressure system (90 psi)
Components are lighter and less expensive
Air is cheap

Hydraulic System

Uses liquid under pressure
Is a high power system
Transmitting medium is incompressible
Leaks can create housekeeping problems
Is a high pressure system (10,000+ psi)

Components are heavier and more expensive
Hydraulic oil is very expensive

Pneumatic systems have their place in low power applications. Before getting into the components of pneumatic systems, let's discuss some facts about air, temperature, humidity, and pressure.

AIR

Air is generally composed of 21% oxygen, 78% nitrogen, and 1% inert gases. The atmosphere at any time can contain 4% water vapor. Since gas molecules are at a distance from each other they can be compressed. The air always assumes the shape of its containers and exerts a pressure at sea level of 14.7 psia.

The quality of air that troubles pneumatic systems the most is humidity. Absolute humidity is the amount of water vapor that the air can carry. This is directly affected by the temperature of the air. Relative humidity is the amount of water vapor carried by a volume of air compared to the amount of water vapor that it could carry (expressed as a percentage). As the temperature of the air increases, the amount of water vapor that it can hold increases. The amount of water vapor the air carries doubles for every 20 degree temperature rise. When the water vapor reaches 100% saturation, the vapor is released as a liquid.

Dew point is the temperature at which the air becomes saturated with water vapor. This is important to pneumatic systems, for when warm compressed air travels through the branch lines it cools. As it drops below the dew point, the water condenses in the lines. The water then damages the air lines (by rust) and the pneumatic operating devices. A point to remember is that if you compress a gas its temperature rises. If you release it from compression it cools.

The flow in a pneumatic system is measured in cubic feet per minute (cfm).

$$1 \text{ cfm} = 7.48 \text{ gpm}$$

A rule to remember in a pneumatic system is that flow causes movement, pressure causes force. To solve for horsepower in a pneumatic system:

$$hp = \frac{flow(cfm) \times psi}{12{,}820}$$

COMPRESSOR

Pneumatic systems must have a prime mover, some sort of power device to power the compressor. The compressor is the start of the pneumatic system. It takes the power from the prime mover and converts it to flow against pressure. The compressor is like a pump; it has an inlet and an outlet.

Inlet Filter

The inlet of the compressor has to be filtered to remove dirt and other contaminants. The inlet filter (Fig. 13-1) may be of two types: the *wet* or the *dry*. The dry is usually a cartridge filter that can be disposed of when it gets dirty. Some dry inlet filters are made to be cleaned and then dried and reused. The wet filters are usually composed of a mesh and a shallow oil reservoir. The air is drawn through the reservoir, and the dust and dirt is dampened by the oil and removed by the filter before the air gets to the compressor. The inlet should always be located in a cool dry area. The cool air molecules make the compressor more efficient, because cool air molecules are closer together. Care should be taken to ensure that no combustible fumes are drawn into the compressor.

Types of Compressors

Compressors can be classified as either positive displacement or nonpositive displacement. *Nonpositive displacement* compressors are usually blowers, being high volume, low pressure. Most pneumatic systems use *positive displacement compressors*. The three main types of positive displacement compressors are the *piston, diaphragm,* and *vane compressors* (Fig. 13-2).

Staging the Compressor

By staging a compressor you can get a boost in the pressure. Staging is a procedure in which one stage of the compressor is

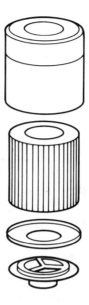

Figure 13-1. Inlet filter.

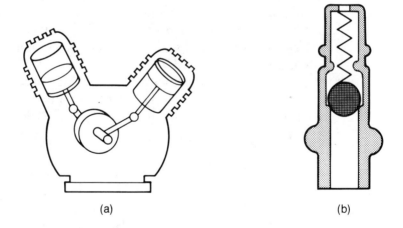

(a)   (b)

Figure 13-2. (a) Compressor; (b) relief valve to prevent excessive pressure build-up in a compressor.

fed into the second stage. The air is already compressed, coming from the first stage, and the second stage compresses it and further increases the pressure. The problem is that when the air is compressed it's hot coming from the first stage.

Intercooler

To cool the air before it gets to the second stage, you may find an intercooler (Fig. 13-3). This cools the air in between stages. By doing this it makes the second stage more efficient by letting it compress cooler air. The intercooler may also remove some condensed water vapor from the air. There are two basic types of intercoolers, air cooled and water cooled. The *air cooled* may be circulating the compressed air through lengths of finned tubing. The *water cooled* (Fig. 13-4) usually circulates the compressed air through a tank containing water lines, through which cooled water is circulated. The water cooled is the most efficient, and is usually equipped with drains to remove condensation.

AFTERCOOLER

After the compressed air leaves the compressor it may be circulated through another cooler called an aftercooler. The aftercoolers are larger than the intercoolers, because they have the air circulating through at a higher pressure. The aftercoolers are usually the water type because they are more efficient than the air cooled type. The aftercooler may have a drain to remove any condensation or the air may pass through a separator (Fig. 13-5) immediately after leaving the aftercooler. The dew point of the compressed air becomes critical at this point. If the moisture isn't removed, as the air is cooled further downstream more moisture may condense and damage the downstream components.

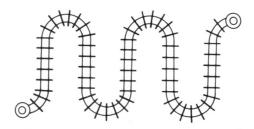

Figure 13-3. Intercooler. The intercooler may be nothing more than finned tubing.

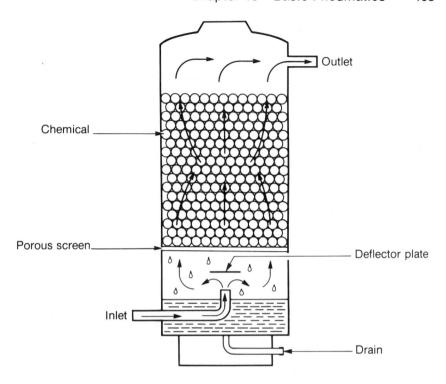

Figure 13-7. Chemical dryer.

the water vapor from the air. As the chemical does so it melts, forming a compound that must be periodically removed. The dryer must also be periodically refilled with the chemical. The *nondeliquescent* (or absorption) dryer uses a chemical that traps the water vapor, and must be periodically dried out for reuse.

Some styles use electric elements in the dryer to heat the chemical to remove the water vapor. Others use hot air blown through the dryer to remove the moisture. These are usually a two-unit type so that one can be drying out while the other is drying the air.

RECEIVER

Next the air is ready to go to the receiver (Fig. 13-8). The receiver is like an accumulator. It stores compressed air under pressure and serves three purposes:

1. Dissipates heat
2. Collects moisture as the air cools
3. Dampens the compressor pulsations and makes the flow smoother and quieter.

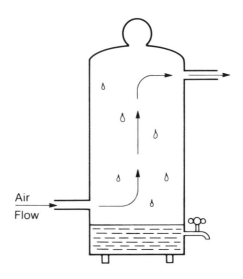

Figure 13-8. Receiver.

FILTER

The next component in a pneumatic system is the filter (Fig. 13-9). Filters remove the contaminants in a system before they get to the downstream components. There are basically two types of contaminants in a pneumatic system:

1. Those generated by the system: rust particles, pipe compound, grit, oil, or water sludge.

2. Those from an outside source: rivets, small nails, metal chips, fine airborne particles, or chemical fumes that the compressor may have sucked into the system.

Some of the contaminants, such as rust (formed by condensation in the lines), form a sludge that will damage the downstream components. This can also restrict air flow, or possibly damage a finished product to which air is being applied (paint, etc.).

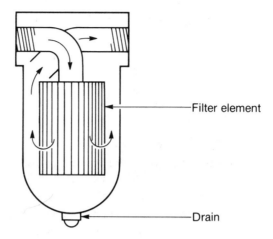

Figure 13-9. Filter.

Filter Ratings

Filters are rated in microns (.000039 inch.). They're given either a nominal rating or an absolute rating. The nominal rated filter will remove the largest percentage of the stated micron size. The absolute rated filter is supposed to remove 100% of its stated size. In actuality the filter removes only about 98%; the other 2% are long, slim, well rounded particles that are not easily trapped.

Types of Filters

Filter elements are made of either cloth or felt or are of wire mesh construction. Cloth filters have high filtration qualities and a low pressure drop. They're usually rated at 5 microns or greater. They can load with water and sludge and deteriorate in the system.

Wire mesh filters are cleanable. They're rated at 40 microns and above. They can clog and create a substantial pressure drop. They can also be damaged very easily during cleaning.

Wire mesh filters should be used on line sizes of 2 inches to ¼ inch. If used on larger lines they aren't capable of allowing the necessary flow. When the filters get dirty, they can create a substantial pressure drop, with decreased flow, because there is no bypass of the filter. The filter can actually stop all flow.

There are also absorption filters designed to collect vapors other than water vapor. These are usually some type of carbon powder or granular chemicals. They're usually rated in the .5 (and above) micron range.

REGULATOR

The next component is the regulator (Fig. 13-10(a)), which controls the downstream pressure. There are several basic types of regulators. The diaphragm type uses a flexible diaphragm to move the valve. As the downstream pressure increases, the diaphragm flexes and closes the valve, not allowing any more flow to pass through. When the pressure decreases, the valve opens and allows flow to continue downstream.

The externally controlled type uses a pilot device downstream to regulate the valve to control the downstream pressure.

Some types of regulators have built-in relief valves to relieve pressure. This is a safety factor when the high pressure could damage downstream components. These valves shouldn't take the place of the standard relief valve found in the vicinity of the compressor.

LUBRICATOR

The lubricator (Fig. 13-10(b)) is found next in the system. An air lubricator injects oil into the air stream to lubricate the internal workings of pneumatic devices. It usually stores lubricant and injects it into the air stream by the venturi effect. The air then carries minute droplets of oil downstream. The problem develops at bends in the lines. They cause the oil to accumulate and to travel as a liquid. This is why the lubricator should be installed as close to the pneumatic device as possible. Another point to consider is that the lubricator can act as a check valve when installed in reverse. It'll restrict flow in the returning line. It makes a big difference how the lubricator is installed in lines that must carry return flow. The regulator has a pressure drop of 2 to 5 psi. To insure proper lubrication it is advisable to use an sae 10 lightweight oil. Light oils are easier to atomize. The compatibility with the lubricator is a factor to watch. Some oils have chemical reactions with the plastic bowls on the lubricator.

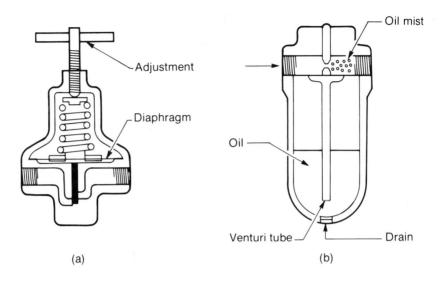

Figure 13-10. (a) Regulator, (b) lubricator.

FRL System

Some pneumatic systems have what is known as an **FRL** system. This is a combination unit of a filter, regulator, and lubricator. The components must be installed in this order to function correctly. The contaminants must be removed, the pressure regulated, then the oil added. If the oil was added first, it would be removed in the filter.

DIRECTIONAL CONTROL VALVE

The next component in the system is the directional control valve. Directional control valves are used to direct or prevent flow through selected passages. There are several styles of directional control valves and each one has its own advantage in certain applications.

Directional control valves can be actuated by several different methods. The following are a list of some examples (also see Fig. 13-11):

- Manual
- Push-pull lever
- Pedal
- Push button
- Mechanical (cam)
- Electric solenoid
- Pilot pressure

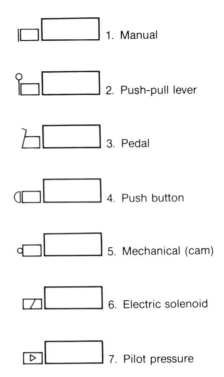

1. Manual

2. Push-pull lever

3. Pedal

4. Push button

5. Mechanical (cam)

6. Electric solenoid

7. Pilot pressure

Figure 13-11. Directional control valve actuators.

There are several varieties of directional control valves for pneumatic systems. Some of the most popular are (1) the two-way valve—an on or off valve; (2) the three-way, which has three flow paths and ports pressure to one end and relieves the same end; (3) the four-way valve which allows four flow paths through the valve; (4) the 5-port valve which comes in three popular types, center exhaust, all ports blocked center, and both cylinder ports open to pressure center (Fig. 13-12).

Each type has its own advantages and disadvantages for use in a pneumatic system. The *center exhaust* allows the actuator to be moved with the valve centered. *All ports blocked center* will not let the actuator move when in the center position (except for the compressibility of the air). The *both ports open to pressure center* holds the actuator in a more rigid position; however, a cylinder has a tendency to move or creep toward the

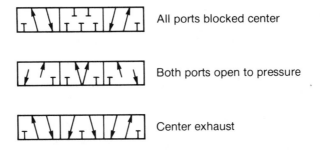

All ports blocked center

Both ports open to pressure

Center exhaust

Figure 13-12. Three common center positions of the 5-port valve.

rod end of the cylinder because of the difference in the areas the pressure is applied.

When looking at directional control valves, the number of positions the valve can be found in is determined on the print by the number of envelopes found in the valve diagrams.

FLOW CONTROL VALVE

The flow control valve (Fig. 13-13) is found next in the circuit. The flow control valve is used to control the flow into an actuator. This then controls the speed at which the actuator works. There are two basic types of flow controls, the *adjustable* and the *adjustable with by-pass*. The speed of the actuator can be controlled by allowing more or less flow to the actuator. The by-pass allows free flow in one direction, making speed control in one direction possible. If the flow is controlled going into the cylinder it is a *meter-in* circuit. If it is controlled coming out of the actuator then it is a *meter-out* circuit.

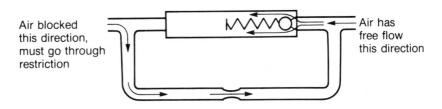

Air blocked this direction, must go through restriction

Air has free flow this direction

Figure 13-13. Flow control valve.

ACTUATOR

The actuator is the last component in the system. There are two types of actuators, cylinders or motors (Fig. 13-14).

Cylinder Actuator

Cylinders provide linear thrust and straight line motion. There are several types of cylinders. The *single acting cylinders* apply force in one direction and are spring, weight, or gravity return. The *double acting cylinders* apply force in both directions, which means it can be driven in both ways. The *double rod cylinder* applies an equal force in both directions. There are also *cushioned cylinders*, which slow down at the end of their stroke to prevent shock to the system. The cylinder retracts until the cushion comes in contact with the opening in the end on the cylinder. The air is no longer able to exhaust out of the cylinder unrestricted. The flow is controlled by a small valve. This meters out the air to provide the cushion to slow it down. There are two types, either adjustable or nonadjustable.

Air Motor Actuator

The air motor is the other type of common actuator. They are usually piston or vane motors. They work just the opposite of the compressors; the pressurized flow comes in, activates the motor, then is exhausted out. They come in two styles, reversible or nonreversible.

MUFFLER

One last device that may be found at the actuator or at a directional control valve is the muffler (Fig. 13-15). It reduces the amount of noise created when the air is exhausted to the atmosphere. It has some restriction to flow, so in use there'll be some back pressure. Whether this is objectional to your system is an important consideration before using a muffler.

SUMMARY

Pneumatic systems will never replace hydraulic systems in industry. They have their own place. The speed and low power

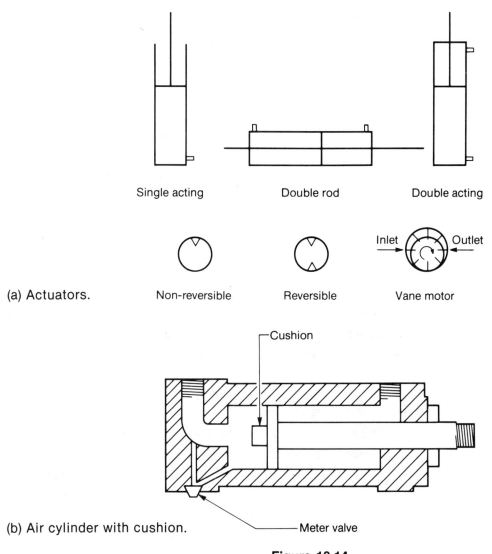

Single acting   Double rod   Double acting

(a) Actuators.   Non-reversible   Reversible   Vane motor

Cushion

(b) Air cylinder with cushion.   Meter valve

Figure 13-14.

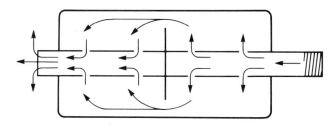

Figure 13-15. Cutaway of a muffler.

they offer make them indispensable in industry. If they're properly cared for they'll give many years of dependable service.

REVIEW QUESTIONS

1. What advantages do pneumatic systems have over hydraulic systems?

2. What problems does the humidity in the air cause in a pneumatic system?

3. Why should the air be filtered before it enters the compressor?

4. What is the advantage of having a multistage compressor?

5. What is an intercooler?

6. What is an aftercooler?

7. What is the purpose of a separator?

8. Why would a chemical dryer be used in a pnuematic system?

9. What three purposes does a receiver serve in the pneumatic system?

10. What is an FRL and what is its purpose?

11. Describe several ways a DCV may be actuated.

12. How does a cylinder cushion work in a pneumatic cylinder?

Chapter 14

Packings and Seals

Pumps, cylinders, and motors of fluid systems must have clearances for moving parts to avoid rubbing on one another. These clearances, however, allow the fluid to leak out (Fig. 14-1). To prevent this leakage, manufacturers have developed seals. Seals are devices for controlling the movement of fluids across a joint or opening in a vessel or assembly.

STUFFING BOX

Packings are a form of seal used where relative motion occurs. The stuffing box (Fig. 14-2) is used to control leakage at a point where a rod or shaft enters an enclosed space that's at pressure above or below that of the surrounding area. The stuffing box has three basic parts:

1. Packing chamber
2. Packing rings
3. Gland follower

The packing is compressed by the gland follower and is forced against the bore of the box and the rotating shaft or rod. The packing must have the ability to deform in order to seal correct-

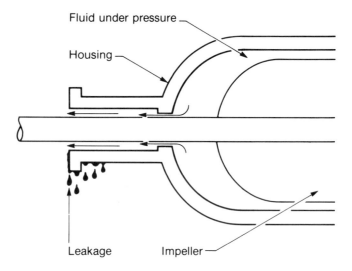

Figure 14-1. Leak.

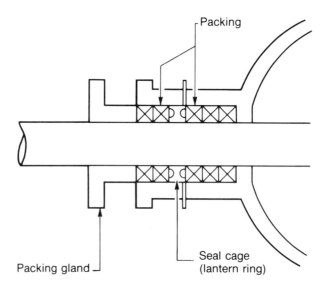

Figure 14-2. Stuffing box assembly.

ly. This type of packing is usually called *compression packing*. There must be frequent adjustments to compensate for wear. The packing must allow some leakage for lubricating purposes. This fluid lost as leakage helps dissipate generated heat and also prevents rapid wear. By not over tightening the packing, it will allow slight leakage.

Packing Materials

Packing materials have four basic styles of construction:

Twisted packing. Twisted packing (Fig. 14-3) is the most widely used and simply constructed. It's not, however, the strongest packing material. It's comprised of asbestos or cotton, lubricated with mineral oil and graphite. Its size is adjustable by merely removing strands.

Square braid. The square braid (Fig. 14-4) is made of many materials (asbestos, cotton, plastic, or leather, and may include metal wires of lead and copper) and usually has eight strands. It's usually grease or oil impregnated. It's very flexible and easily adjusted.

Braid over braid. The braid over braid (or jacket over jacket) (Fig. 14-5) is made of a series of round tubes one over another. It's fabricated of various fibers and impregnated with lubricant. It may be braided over a lead core to help it hold its shape.

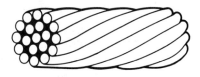

Figure 14-3. Twisted packing.

Figure 14-4. Square braid.

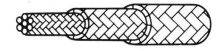

Figure 14-5. Braid over braid or jacket over jacket.

Interlocking braid. The interlocking braid (Fig. 14-6) is a combination of the braid over braid and the square braid. All the yarns are interlocking and have great resistance to unraveling. It's available in asbestos, cotton, plastic, and other fibers and is usually oil lubricated. It's the strongest of the packing materials.

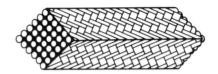

Figure 14-6. Interlocking braid.

Installation of Packing

In looking at the stuffing box, we can get an idea of the reasons for correct installation procedures. The purpose of the multiple rings in a stuffing box is to break down the pressure of the fluid being sealed, so that when the pressure gets to the gland followers it'll be a zero psig.

In practice the bottom ring does most of the sealing. It's important to install it correctly. The common arrangement is the five packing rings and one lantern ring (or seal cage) (Fig. 14-2). The seal cage allows the introduction of a lubricating fluid between the packing and the shaft. It's possible to connect the discharge from the pump to the seal cage to introduce the lubricating fluid to the packing area. In abrasive conditions, or when you're pumping abrasive liquids, it's advisable to connect a line of cool clean water to the seal cage.

When packing a stuffing box with a lantern ring it's important to be sure that the seal cage is in line with the inlet port. As the packing is periodically adjusted for wear, the seal cage moves to the back of the stuffing box. The seal cage must be positioned so that it's always under the inlet port. It may be troublesome, but the life of the installation will be severely shortened if the seal cage is left out.

The following is a ten-step procedure for correctly installing packing:

1. Remove all old packing and clean the stuffing box. Make sure the inlet port to the seal cage is open and clean.

2. Cut the packing rings on the pump shaft or one of the same size (Fig. 14-7).

3. Use the butt joint on all cuts except for valves and expansion joints (then use skive cut 45 degrees).

4. Put the first ring in, being careful to get it fully seated in the bottom. You may want to use an appropriate sized bushing or sleeve to assist.

5. Insert additional rings, being careful to stagger the joints for proper sealing and support.

6. Properly position the seal cage.

7. Finish installing the additional rings.

8. Install the gland follower. It should extend about 1/3 the depth of the packing.

9. Tighten the follower as you rotate the shaft. This prevents the follower from cocking and binding.

10. Open the valves and start the pump. Inspect for the proper amount of leakage. Make any necessary adjustments. Recheck in a few hours; additional adjustments may be required.

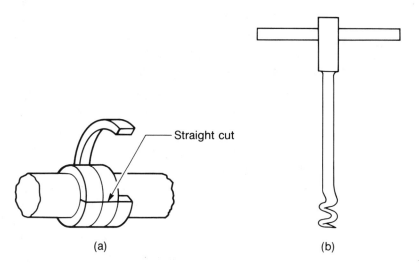

Figure 14-7. (a) Cutting packing on a shaft, (b) tool to remove old packing.

MECHANICAL SEALS

In recent years, pump manufacturers have been installing a different type of seal on pumps called mechanical seals. These are much more expensive than the packing installed in the stuffing box installation, but they have no leakage. In some industries, leakage from stuffing boxes is unacceptable. The mechanical seal eliminates this problem. The mechanical seal (Fig. 14-8) has five basic parts:

1. Rotating member
2. Stationary member
3. Set screws
4. Spring loaded collar
5. Elastomer

These five parts make up the standard mechanical seal.

Types of Mechanical Seals

There are three types of mechanical seals.

Inside seal. The inside seal (Fig. 14-8) fits inside the stuffing box. The fluid pressure inside the stuffing box helps to hold the faces of the seal together. If the fluid pressure is excessive, a balanced seal may have to be used. A balanced seal has a cutaway on the inside of the seal to help balance out the pressure. The contact surfaces of the mechanical seal must never run dry. There must be some fluid in the pump; if there isn't, the friction will generate enough heat to destroy the seal.

Outside seal. The outside seal (Fig. 14-9) is located on the outside of the stuffing box. Since no rotating parts of the seal are located inside the stuffing box, it's a good seal to use for sealing corrosive and abrasive materials.

Double mechanical seal. The double mechanical seal (Fig. 14-10) is two single seals back to back. This seal is usually used for hazardous liquids. A clear fluid is circulated through the seals at higher than system pressure. This liquid would leak inward instead of outward, thereby assisting the seal to prevent leakage of the hazardous material.

All three of the seals must have some lubrication of the con-

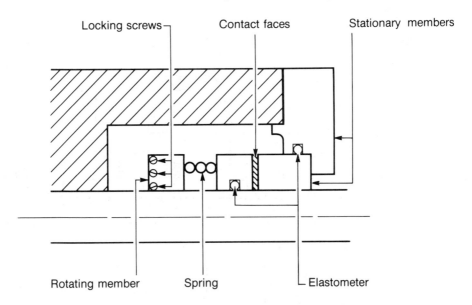

Figure 14-8. Inside mechanical seal.

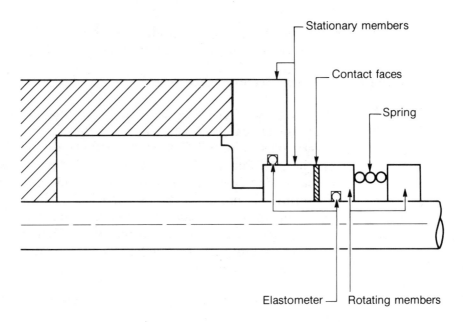

Figure 14-9. Outside seal.

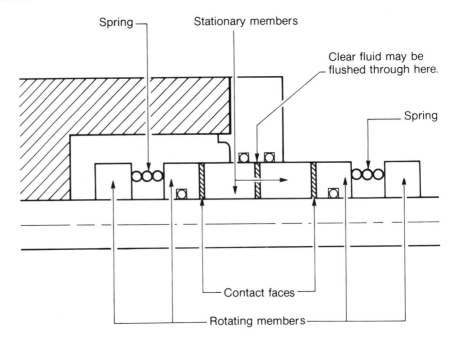

Figure 14-10. Double seal.

tacting surfaces by a liquid. Running any of them dry will destroy them in a matter of seconds.

Installation of a Mechanical Seal

No one method or procedure can be outlined for installation of a mechanical seal. Some can be assembled in only one way and are very easy to install. Some may be installed only after inspecting the location and determining which procedure is correct. The outside is easy and obvious. The inside is harder because the equipment must be disassembled. The best method is to follow the procedure that's packed inside each seal by the manufacturer. Some manufacturers make seals that are preadjusted and merely have to be bolted on the pump.

Some installation points to watch are:

1. Check the shaft for run-out and endplay (maximum of .005).

2. Clean all burrs and sharp edges.

3. Be sure there are no nicks or scratches in seal faces.

4. Don't allow faces to make dry contact.

5. Lubricate the seal faces with oil or the fluid to be sealed.

6. Protect all static seals from sharp edges.

7. Make sure the seal is surrounded by the liquid before start up.

O-RING PACKING

The o-ring is a squeeze type packing made from synthetic rubber or similar materials. The most common shape is the circular cross section. The principle behind the o-ring seals (Fig. 14-11) is called controlled deformation. A slight squeeze puts the o-ring into contact with both surfaces. The compression keeps the surfaces in contact. Additional deformation is caused by the pressure the fluid exerts on the o-ring.

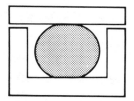

Figure 14-11. O-ring.

The o-ring is usually used for static sealing and reciprocating motion. It can be used for oscillating and rotary motion if the speed is kept low. The o-ring usually fits into a rectangular groove that's $1\frac{1}{2}$ to 2 times the width of the o-ring. This allows the o-ring to slide and roll in the groove (Fig. 14-12). The reason for this is that first, it distributes wear on the o-ring, and second, it helps to lubricate the sliding surfaces. O-rings require about 10% initial preload to work properly. They're usually manufactured so that when they're installed they have the necessary preload. There's presently no uniform sizing code in industry. Sizing varies from manufacturer to manufacturer.

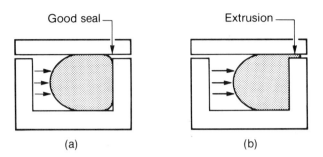

Figure 14-12. O-ring deformation.

LIP PACKING

In addition to o-rings, industry uses a formed and molded packing to help eliminate leaks. One of the most common styles is lip type packing. Lip packing must be installed so that the lips of the packing expand and collapse properly. If they're over tightened, they're improperly preloaded and become compression packings. A slight preload is required, but the sealing should occur as a result of the size and shape of the packing.

Types of Lip Packing

Lip type packings come in four basic types.

Cup packing. Cup packing (Fig. 14-13) is one of the most widely used types of packing. It's highly satisfactory for plunger end applications. It's an unbalanced packing since it has only one lip. The inside follower must only be snugged to prevent initial leaking. The clearance between the back plate and the cylinder walls must be very small. The clearance between the packing and the follower must be enough to allow the lips freedom to work.

U-packing. U-packing (Fig. 14-14) is a balanced packing because it has two lips. It seals on the inside and outside surfaces. It must have some sort of support to prevent the lips from collapsing. For support, it usually uses some sort of material (flax, rubber, hemp, fiber, etc.) as a filler. It may even be supported by a metal ring (commonly called a pedestal ring). The ring must be such that it allows the lips freedom of movement to work properly.

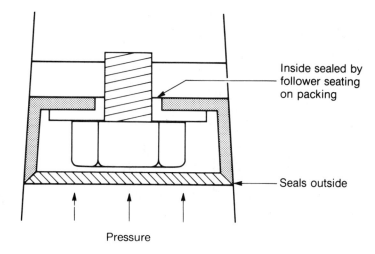

Figure 14-13. Cup packing.

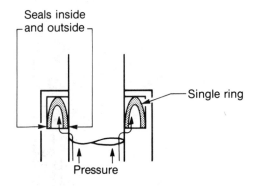

Figure 14-14. U-packing.

Flange packing.   Flange packing (Fig. 14-15) is the least used packing because it's good only for low pressure since it seals only on the inside surface. The outside surface must be sealed by the clamping force.

V-shaped packing.   V-shaped packing (Fig. 14-16) is the most popular packing. It's effective on high or low pressure, and on rotating or reciprocating applications. The inside angle is a standard 90 degrees. It has the advantage that after it's worn slightly, it can be tightened to seal further. It's usually used in multiple packing. It has a support and adaptor ring at the top

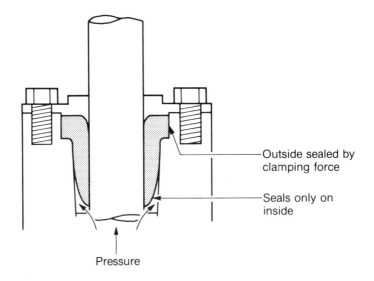

Figure 14-15. Flange packing.

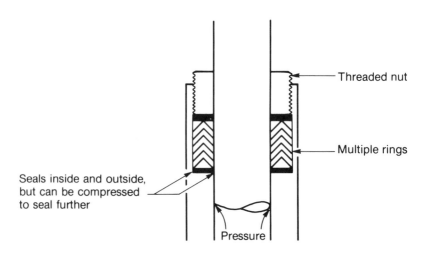

Figure 14-16. V-shaped packing.

and the bottom. The support ring is usually metal and the adaptor ring is made out of some other type of material.

Some points to consider when working with the lip type packing are as follows:

1. Understand how the packing is to work. If installed incorrectly, it may trap fluid under pressure and thus wear the packing out much faster than it should.

2. Remove the old packing and examine it. Look for clues as to why it failed.

3. Look for worn surfaces at the end of the stroke or in reversal areas.

4. Check for dirt; clean the system if necessary.

5. Check the new packing for the correct size.

6. Make sure you have the correct adaptors, support rings, fillers, etc.

7. Look at the assembly and understand how it works.

8. Install correctly and do not overtighten.

SUMMARY

A proper understanding of packing and sealing is important to prevent leakage. In some industries, leakage represents a loss of money, in others a hazard. No matter which problem faces the repairman, there is some form of sealing device to solve the problem. It is up to the repairman to decide which method is most suited to each case.

REVIEW QUESTIONS

1. Describe the basic construction of a stuffing box assembly.

2. Describe the four basic types of packing.

3. What purpose does a seal cage serve in a stuffing box assembly?

4. List the steps involved in properly repacking a stuffing box.

5. What are mechanical seals and how do they seal?

6. What are the three types of mechanical seals and the purpose of each?

7. What is an o-ring and how does it seal?

8. What are the four basic types of lip packing?

9. What are eight points to consider when installing lip packing?

Chapter 15

Variable Speed Drives

Gear, chain, and belt drives are fixed speed drives. To change the output speed requires a mechanical change. Variable speed drives are designed to overcome this problem of changing output speed. They can be adjusted so that for one input speed, there can be a range of output speeds. There are three basic types of variable speed drives: mechanical, hydraulic, and electrical.

MECHANICAL VARIABLE SPEED DRIVES

Mechanical variable speed drives are divided into four basic categories: Open belted, enclosed belted, chain type, and traction drives.

Open Belted Drives

Open belted drives (Fig. 15-1) are composed of a motor and two pulleys and a variable speed belt. The motor usually has the adjustable pulley mounted on it, and the spring loaded pulley is mounted on the output. As adjustment is made, the size of the motor pulley can be increased until the belt rides high on the pulley. This makes the belt ride lower in the spring loaded

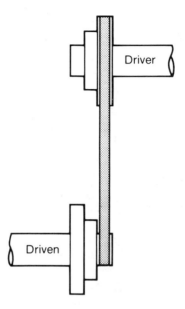

Figure 15-1. Open belted drive (step pulleys). The drive ratio is determined by which set of pulleys the belt runs on.

pulley (Fig. 15-2). This makes the output run at a faster speed than the input.

If the motor pulley is adjusted till the belt rides almost at the bottom of the pulley, the belt will ride close to the top of the spring loaded pulley. This makes the output run slower than the input.

This type of drive can also be accomplished by using a movable motor base and a spring loaded pulley. By adjusting the motor base, it adjusts where the belt rides in the spring adjustable pulley.

Enclosed Belted Drives

Enclosed belted drives (Fig. 15-3) work on the same principle as the open belted, the difference being that the motor, the drive, and usually a small geardrive are all contained in one unit. The adjustment is usually made by a hand crank or can be remote controlled from a distance by flexible shafting, or even a small motor drive.

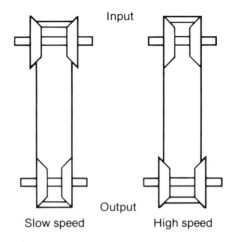

Figure 15-2. Pulley-belt positions.

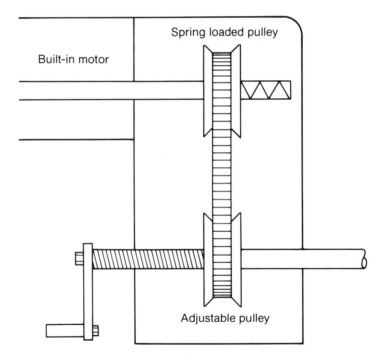

Figure 15-3. Enclosed belted drive.

Chain Drives

Chain type drives (Fig. 15-4) are composed of a chain made up of many slats, which are engaged with two pulleys that have ridges running radially. The engagement between the two is a positive type with no slippage. The drive has one spring loaded pulley and the other is mechanically adjustable. The chain is adjusted to ride in different positions on the grooved wheels. The difference in speed can be adjusted from high output to low output depending on the position of the chain.

In this drive the amount and type of lubricant is very important. Too little or the wrong type can be fatal to the drive. It's best to obtain the manufacturer's recommendation for each type of drive.

Figure 15-4. P.I.V. type drive. (Courtesy of P. T. Components, Inc.)

Traction Drives

Traction type drives use metal-to-metal contact of various components to transmit power. The most common types of traction drives are the ring roller, the ball type, and the cone type.

Ring roller drive.   The ring roller transmission (Fig. 15-5) has a series of cone shaped rollers that have a stationary ring holding them in a fixed position. The input is a series of

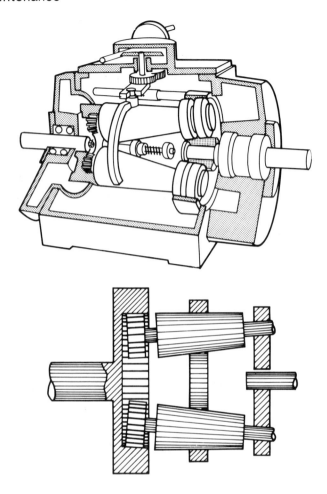

Figure 15-5. Ring roller.

planetary gears fastened to the rollers. The output is a sliding mechanism that adjusts to the running circumference of the rollers. By adjusting the sliding holding ring, the outer diameter of the output rollers is adjusted. When the output circumference is large the output speed is slow. When the output circumference is small, the output speed is faster. This unit has the advantage of being able to be adjusted either while stationary or while running.

Ball type drive.   The ball type transmission (Fig. 15-6) uses a

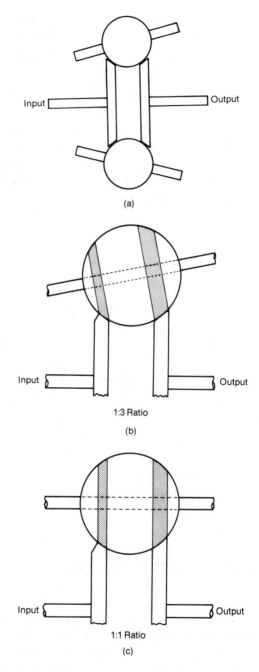

(a)

1:3 Ratio

(b)

1:1 Ratio

(c)

Figure 15-6. Ball type transmission. (a) The tilt of the balls determines the speed change. (b) Ball position for 1:3 ratio. (c) Ball position for 1:1 ratio.

ball running between the input and output discs. As the ball is adjusted the rings run on different circumferences on the ball. When the input runs on the small circumference and the output on the large circumference the output speed is high.

Cone type drive. The cone type transmission (Fig. 15-7) consists of a cone driven by the input and disc riding on the face of the cone. The position that the disc rides on the face of the cone determines its output speed. If it rides low on the cone, the speed is faster. If the disc rides high on the cone, its output speed is slower. This style of drive should be adjusted only while running.

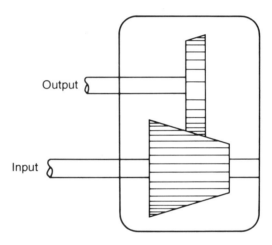

Figure 15-7. Cone type variable speed drive. The position in which the disc rides on the cone determines the output speed.

HYDRAULIC VARIABLE SPEED DRIVES

The most common type of hydraulic variable speed drive is the hydrostatic drive. In basic hydraulics, by varying the output of a pump to an actuator, we can control its speed. The same is true in hydrostatic systems. By controlling the flow, we can control the speed. There are two basic types of hydrostatic drives, the open and the closed.

Open System Hydrostatic Drive

The open system (Fig. 15-8) uses a variable displacement pump or a flow control valve to control the flow to the hydraulic motor. The speed is adjusted by adjusting the flow to the motor.

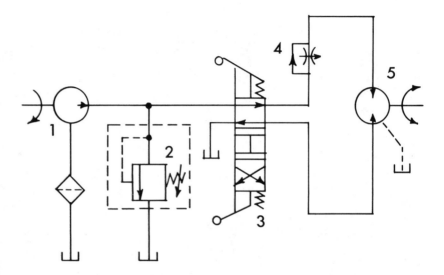

1 Hydraulic pump
2 Pressure relief valve
3 Directional control valve
4 Flow control valve
5 Reversable motor

Figure 15-8. Open system. Speed is varied by adjusting the flow control valve.

Closed System Hydrostatic Drive

The closed system (Fig. 15-9) uses a pump and motor combination, where the discharge from the motor is fed back to the pump. The fluid lost by internal leakage is replenished from a reservoir. In most cases the pump and motor are both reversible to give the drive a reversible feature.

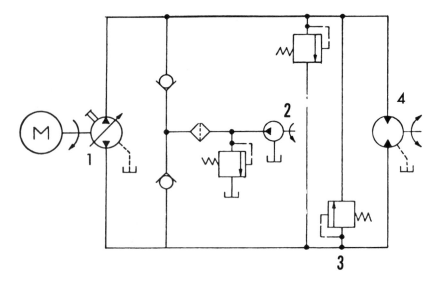

1 Variable displacement pump   3 O.L. relief valve
2 Replenishing pump   4 Fixed displacement motor

Figure 15-9. Closed system. Speed is varied by adjusting the volumetric displacement of the pump.

SUMMARY

In this section we've looked at several types and styles of variable speed drives. The drives are each suited for certain applications. There's no variable speed drive that can perform best in all categories. The situation that the drive is needed for determines which drive can be used. This is usually best done by the original equipment manufacturer.

REVIEW QUESTIONS

1. What are the three basic types of variable speed drives?

2. What are the four basic types of mechanical variable speed drives? Explain the operation of each.

3. What are the two basic types of hydraulic variable speed drives?

Chapter 16

Preventive Maintenance

Preventive maintenance is an important part of every maintenance program. It's essential if equipment is to deliver the service it was designed to provide. The basic objective of good maintenance is to prevent interruptions to the equipment operations. This can be accomplished by good inspection, scheduling, and follow up of a preventive maintenance program.

Preventive maintenance endeavors to eliminate any downtime of equipment. One of the most essential parts of any preventive maintenance programs is choosing the correct personnel to design and implement the program. There are two key people in a preventive maintenance program, the inspector and the scheduler.

INSPECTION

The inspector should be familiar with the equipment. The familiarity with the equipment enables the inspector to know when something is malfunctioning. He should be able to spot most equipment problems before they progress to a point that the equipment breaks down. His inspections should be turned in to a scheduler, who in turn will schedule the repairs necessary when they won't interfere with production. This could be on a scheduled downturn, a nonoperating turn, or any period when the equipment is not being used. The inspector should note on his report how long a period of time (in his estimation) that the equipment can continue to run before the repairs must be made. This is important for it gives the scheduler a time frame to work in. Without this estimation, the scheduler may wait too long before scheduling the repairs, and the equipment may break down.

SCHEDULED MAINTENANCE

In addition to inspections, scheduled maintenance is important. If any equipment is to remain productive, there must be a routine maintenance program. This routine (or scheduled) program should consist of certain checks and services that should be performed at specific time intervals. It might be compared to the service intervals on an automobile. These services usually fall into five categories: daily, weekly, monthly, semiannually, and annually. The scheduler must pay attention not only to the inspections but also the time frame on these routine items.

Basic routine items and their service intervals are recommended by the manufacturer of the equipment. It's best to follow his recommendations when setting routine service intervals. If in doubt, contact the manufacturer. They're very considerate and want their equipment to provide the best service possible.

The scheduler should be someone who is familiar with both the maintenance repair work and also the production schedule. This will enable him to best fulfill his assignments. The scheduler is responsible for scheduling the following items:

1. Inspections of equipment.
2. Routine maintenance of equipment.
3. Repair assignments to the millwrights.
4. Scheduling the follow up inspections of the performed work.

It's beneficial if standard forms are provided for the inspections. This makes it easier for all involved.

In addition to these scheduling assignments, the scheduler needs to keep a record of all breakdown repairs. This will enable the maintenance supervision to keep track of all problem areas. If a certain piece of equipment has an abnormal number of breakdowns, the supervisor may change some aspect of the routine maintenance performed to eliminate the problem. Detailed inspection of all equipment failures to determine the cause will help eliminate continued breakage of the same component. The inspector can be used to do this type of inspecting. This takes considerable practice, but once the individual becomes proficient at this type of inspection he can save considerable time and money spent on repetitive repairs.

NONDESTRUCTIVE TESTING

Nondestructive testing is another method of inspection that's becoming prevalent in industry. This is usually divided into four basic categories: particle dye, ultrasound, vibration analysis, and sample testing.

Particle Dye Test

Particle dye is used to check for defects in equipment. It usually consists of a magnetic dye and a powerful magnet. The dye is spread on the piece of equipment, the magnet is turned on (or placed on the equipment), and the excess is brushed away. Any cracks or defects in the equipment will draw the dye inside. Then some type of light (usually ultraviolet) shows the inspector the location of the cracks or defects. The repair or replacement can then be recommended.

Ultrasound Testing

Ultrasound uses sonic waves and a monitor to spot defects in any material. This is also a useful tool in spotting subsurface defects in equipment without costly disassembly or surprising breakdowns.

Vibration Analysis

Vibration analysis uses a vibration monitor to determine if defects are developing in equipment. The analyzer usually displays waves on a screen according to the type of vibration it senses. By using charts, the inspector can pinpoint what's causing the vibration, and can make recommendations that can eliminate the vibration.

Another type of analyzer uses a transmitter on the equipment that you plug a meter into. The meter gives a numerical output on its display. The repairman reads the display, compares it to the chart, and can tell what the condition is of the particular component. If accurate records are kept, the equipment's gradual deterioration can be charted and a schedule can be set up for replacement of components. This will help prevent unsuspected equipment breakdowns.

Sample Analysis

Sample analysis consists of taking oil samples from drive systems, hydraulic, or pneumatic systems, and analyzing them to pinpoint any wear in the system. This can be done by part of the maintenance department, or if the equipment is not available by an outside company. (There are presently several that specialize in this type of testing.) This is an effective method of discovering defective components before they fail during production.

All testing equipment is very important to the inspector. If he can't spot potential problems during his inspections, the maintenance program will not function properly. If good communication is lacking between the inspector and the scheduler the program will suffer. If the inspections and scheduling are both performed correctly, the benefits will be self-evident.

Preventive maintenance is becoming so important that many industrial plants are investing in computerized scheduling. The computer allows the scheduler to be more flexible and also keeps more accurate records. In addition, all costs—both personnel and material—can be kept for all equipment.

If industry is to be productive, preventive maintenance is very important. If maintenance costs are to be reduced and production improved, equipment will have to be serviced regularly. If not, the results may prove detrimental to the economy of the company.

REVIEW QUESTIONS

1. What is the function of the inspector in a preventive maintenance program?

2. What is the function of the scheduler in a preventive maintenance program?

3. Why is it important for the inspector to make a note of the time required for the repair?

4. Why is it important for the inspector to make note of the seriousness of the repair needed?

5. What device are modern companies using to monitor their preventive maintenance programs?

Chapter 17

Mechanical Troubleshooting

To become a good repairman it's essential to be a good troubleshooter. The following material is intended to serve as a general guide to mechanical troubleshooting. It won't cover all situations, but may be adapted to most situations.

To troubleshoot mechanical drives it's important to break the equipment into its basic components. This chapter is divided into bearings, belts, chains, and gears.

BEARINGS

Bearing troubleshooting is a combination of using your senses and knowledge to identify problems. Bearing problems usually show up in the form of heat and sound or vibration.

Heat

Heat can be discovered by merely feeling the housing in which the bearing is mounted. If the housing can be touched and isn't uncomfortable to keep your hand on, the temperature is probably safe. If you can't keep your hand on it more than a second, the temperature is becoming a problem, and should be investigated. If it's so hot it can't be touched at all, then it's critical. At this temperature the grease can oxidize, the steel can change its material structure, and the bearing can expand internally and destroy the running clearance.

That method may seem antique, but many plants don't have temperature monitors installed on their bearings. If they do have monitors on the bearings, the guess work is eliminated. The temperature can be read and any necessary action can be taken.

Noise

Noisy bearings have some type of contamination problem. The contaminant is causing some form of metal-to-metal contact, causing the noisy bearing.

Bearings that are running rough may have contamination, or may have the incorrect shaft and housing fit. The incorrect clearances cause the vibration in the bearing.

Conditions to be Checked

Any bearings that are suspect should be checked as quickly as is practical. Some of the conditions that may be checked are as follows:

1. **Fatigue failure** (Fig. 17-1). This type of failure has two main causes: normal wear and overload on the bearing. Normal wear means the bearing has lived its normal service life and needs to be replaced. Less than 5% of all bearings reach this point. Overload can be of two types: too much load and too much speed. The load can be a parasitic load, which is a load which doesn't belong, but is present due to some neglect on the repairman's part. If the housing is too small or the shaft too large, or if a burr in the shaft or housing is present, the internal dimensions of the bearing are changed. This results in additional loading placed on the bearing. While it doesn't result in immediate failure, it drastically shortens the bearing's life. Every effort must be taken to insure that the operating conditions are as close to perfect as possible.

2. **Contamination** (Fig. 17-2). This occurs when dirt enters the

Figure 17-1. *Fatigue.* The normal failure of a bearing is fatigue. Pictured is a typical fatigue failure on an inner ring. The coarse grained pattern should be noted in contrast to the pattern of a lubrication or abrasion failure. Good loading conditions are evident, the load zone arcs on both roller paths being of equal length. (Courtesy of The Torrington Co.)

Figure 17-2. Contamination. (Courtesy of Fafnir Bearing Div. of Textron, Inc.)

bearing, causing the bearing to wear. As the wear causes the internal geometries of the bearing to change, the bearing becomes noisy and vibration begins. As time progresses, the bearing won't be able to remain in service.

3. **Brinelling** (Fig. 17-3). The blame for this problem almost always rests with the repairman. It's caused by incorrectly applying force to the wrong bearing race while mounting or dismounting the bearing (Fig. 17-4). One rule to remember: WHEN MOUNTING A BEARING, APPLY THE FORCE TO THE TIGHT FIT RING. This will eliminate most of the brinelling problems. Brinelling is actually a dent in the raceway caused by a force applied to the rolling element. The rolling element applies a force to the race exceeding its elastic limit, leaving the dent.

Figure 17-3. *Brinelled raceway.* Brinelling of the raceway illustrated was caused by roller impact. The damage occurred during mounting of the bearing. There is a displacement rather than a loss of metal in brinelling. Raceway surface brinelling results in a noisy bearing and such a mark can be the nucleus for premature failure. (Courtesy of The Torrington Co.)

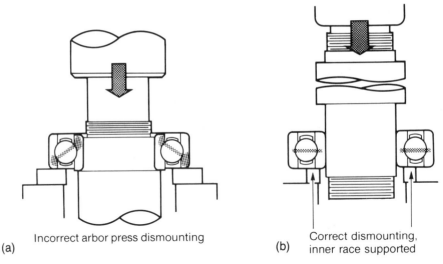

(a)   Incorrect arbor press dismounting

(b)   Correct dismounting, inner race supported

Figure 17-4. Incorrect versus correct method of mounting bearings: (a) force applied to wrong race, (b) force correctly applied, inner race properly supported (Courtesy of The Torrington Co.)

4. **False brinelling** (Fig. 17-5). While this failure resembles true brinelling, it's caused by an entirely different set of circumstances. Brinelling is a dent; false brinelling actually has material removed from the bearing race. Three things must be present to cause false brinelling: a stationary bearing, mounted under load, and an external vibration. The vibration under load, on the stationary bearing, causes the metal-to-metal contact. The vibration causes the ball and the race to work against each other. This wears material away from both parts. False brinelling can be eliminated by removing any one of the three conditions. It will not occur unless all three are present.

5. **Misalignment** (Fig. 17-6). This problem is usually apparant by the path the balls or rollers leave on the raceway. The best way to prevent it is to be sure that all components are properly aligned during installation. Problems as pictured in Fig. 17-7 will cause rapid bearing failures.

6. **Electric arcing and fluting** (Fig. 17-8). This is caused by electric current passing through the bearing. It may be caused by using the bearing for a path to ground while welding. It may also be caused in motors and generators by a breakdown in insulation. Some machinery develops enough static electricity to

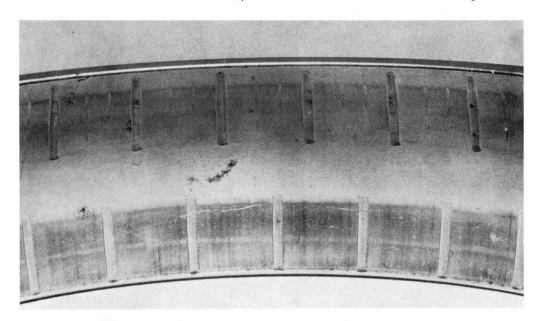

(a)

Figure 17-5. *False brinelling.* While true brinelling is a flow of material due to excessive pressure that causes indentation in a part, false brinelling involves an actual removal of material and is a wear condition. The exact cause of false brinelling is not agreed upon by authorities. It is known that relative motion, load, and oxygen are prerequisites. Other names for this phenomenon are fretting corrosion and friction oxidation. (Courtesy of The Torrington Co.)

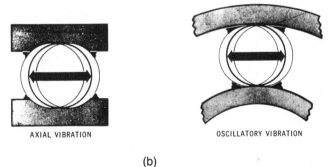

AXIAL VIBRATION   OSCILLATORY VIBRATION

(b)

Figure 17-5. (b) Vibration that causes false brinelling. (Courtesy of Fafnir Bearing Div. of Textron, Inc.)

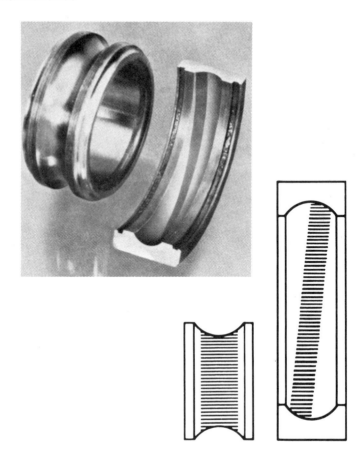

Figure 17-6. Misalignment. (Courtesy of Fafnir Bearing Div. of Textron, Inc.)

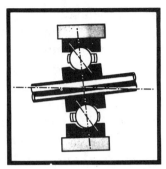 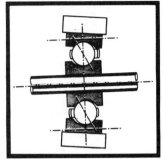

Misalignment of the shaft in relation to the housing causes an overload of the balls which will result in the failure described.

Housing misalignment may be caused either by the housings being cocked with the plane of the shaft or the housing shoulder being ground out-of-square so that it forces the outer ring to cock in relation to the inner. It may also result from settling of the frames or foundations.

(a)   (b)

Figure 17-7. (a) Shaft misalignment, (b) misalignment of the housing.

(a)   (b)

Figure 17-8. Electric arcing (a) and fluting (b). (Courtesy of Fafnir Bearing Div. of Textron, Inc.)

cause this form of damage. When observed, some measure must be taken to prevent current passing through the bearing or eventual premature failure will result.

7. **Lubrication failure** (Fig. 17-9). This type of failure is caused by one of four conditions: too much, too little, wrong type, or dirty lubricant. It's usually observed by a burnt smell and a darkened color to the lubricant. The bearing races may be discolored, and they'll be noisy. The cure is to use the right amount of the correct grade of clean lubricant.

Most bearing failures fall into these categories. If the inspection can find the bearing before it completely fails, the problem is apparent. If not, the bearing may be in such bad shape that no problem is clearly seen to be the cause. At the first sign of abnormal heat, noise, or vibration the bearing should be inspected for cause. Correcting the cause at the earliest possible time pays off in longer bearing life.

Figure 17-9. Lubrication failure. (Courtesy of Fafnir Bearing Div. of Textron, Inc.)

BELT DRIVES

Most belt failure information is covered in Chapter 7, Belt Drives. Included here, however, is a troubleshooting chart for quick access to the information. (Table 17-1)

CHAIN DRIVES

Troubleshooting chain drives is a matter of observation. Table 17-2 is a list of common problems with chain drives.

GEAR DRIVES

The information on tooth failures in Chapter 9, Gears, is very useful in troubleshooting gear drives. When inspecting the teeth, the photographs in that chapter should be used to identify problem areas.

The two most common areas of problems in gear drives is *lubrication breakdown* and *misalignment*. If these two areas are considered in routine maintenance, there'll be very few gear breakdowns to troubleshoot.

Table 17-1   Belt Failure

Problem	Possible Cause	Cure
Abrasion on the belt	Dirt in sheaves	Clean sheaves and remove source of dirt
Belts turn over in sheave	Guard rubbing belt Belts slipping in sheave Damaged tensile section Misalignment of pulleys	Adjust or replace guard Readjust tension Properly install new belts Check alignment, realign if necessary
Belt squeal	Incorrection tension	Check and adjust tension if necessary
Belt swollen and sticky	Belts overloaded Oil or grease on belt	Use higher rated belt Replace belts and eliminate source of oil or grease
Belt broken	Drive overload Foreign matter in sheave	Check drive, use higher rated belt Remove material and check guard, keep foreign matter out of sheaves
Bottom of belt cracking	Belt slipping causing hardening of the belt underside Backside idler	Check tension, adjust to avoid slippage Increase diameter of idler Use of another idler recommended
Belt whip	Insufficient tension A too long center distance	Correct tension Increase tension, use kiss idler
Belt stretch	Broken tensile members	Properly install new belts Check for drive overloads

Table 17-1   continued

Rapid belt failure	Worn pully sheaves	Check sheave for wear
	Belt damaged during installation	Properly install new belt
	Misaligned pulleys	Check and adjust alignment
	Foreign material in sheave	Remove material and improve guard
	High temperatures	Ventilate guard, remove source of heat
Spin burns	Belt slipping during initial start-up	Adjust tension, check belt rating
	Locked sheave	Check to insure all components will turn
Loose belt in a multiple set	Belts mismatched	Check match code
	Old and new belts used in same set	Do not mix old and new belts, change all belts in the drive
Backing cut in a series of banded belts	Worn sheaves	Check sheaves for wear
Banded belts	Belt rubbing an obstruction	Check belt path, check guard

Table 17-2 Chain Drive Problems

Problem	Possible Cause	Cure
Noisy drive	Misalignment	Check sprocket alignment
	Incorrect tension	Check slack side span for 2% deflection
	Insufficient lubrication	Check drive to insure proper lubrication
	Worn drive	Check chain and sprocket for wear
Wear on inside of link plates and sides of sprocket teeth	Misalignment	Check drive for correct alignment
Chain climbs sprocket teeth	Chain worn	Check for chain wear
	Insufficient tension	Check chain tension
	Material in tooth pockets	Clean sprocket, eliminate cause of material buildup
Chain hangs in sprocket	Worn sprocket teeth	Check for sprocket wear
	Sticky lubricant	Use correct grade of lubricant
Stiff chain joints	Misalignment of drive	Check and correct alignment
	Worn and corroded chain	Replace chain and provide correct lubrication
Chain whip	Overloads	Reduce loads, check drive rating
	Insufficient tension	Correct the tension
	Stiff chain joints	Replace bad joints, or install new chain
	Fluctuating loads	Increase chain size, use spring-loaded idler to dampen pulsation
Broken sprocket teeth	Chain climbing sprocket	Check for chain wear, excessive slack
Drive excessively hot	Running too fast	Check drive for proper speed
	Insufficient lubrication	Check for correct lubrication
	Chain rubbing on obstruction	Check drive for obstruction, check guard design
	Wrong type of lubrication system being used	Check drive design for proper lubrication system

SUMMARY

Becoming an effective mechanical troubleshooter takes time and practice. The skill must be developed over a period of time. The tables in this chapter will give a repairman a start in troubleshooting.

REVIEW QUESTIONS

1. What is brinelling and what causes it?

2. What is false brinelling and how is it different from true brinelling?

3. Are all fatigue failures normal failures?

4. What are some common belt failures and how do you prevent them?

5. What are some common chain failures and how do you prevent them?

6. What is the best way to anticipate problems with gears?

Chapter 18

Hydraulic Troubleshooting

Tables 18-1, 18-2, 18-3, 18-4, and 18-5 are divided into sections by components. Usually, trouble in hydraulic systems are identified by certain operating conditions. The charts will prove to be useful in determining hydraulic system problems.

Table 18-1   Hydraulic System Problems

Problem	Possible Cause	Cure
Excessive system heat	System pressure too high	Reduce pressure to correct setting for system
	Pump not unloading during system inactivity	Check unloading valve to insure that its setting is below that of the relief valve
	Cooler stopped up	Check for water flow
	Cooler too small	Check engineering specifications
	Reservoir too small	Check specifications
	Oil level too low	Add oil
	Internal pump leakage	Change pump
	Leakage through system components	Find the component with the hot return line indicating leakage
	Undersized piping in system	Check the original sizes to insure no changes have been made
Excessive wear on system components	Incorrect viscosity oil in system	Check manufacturer's minimum viscosity requirements
	Foreign material circulating through system	Clean the system, change all filters, add new oil
	Air in system	Stop air entry into system, usually on inlet side of pump
Noisy pump	Cavitation	Check inlet pressure, insure that there are no inlet restrictions
	Air leak on inlet side	Be sure all fittings are tight
	Misalignment	Check coupling alignment
	Low oil level	Fill to proper level to prevent the pump from drawing in air

Table 18-1   continued

Noisy pump continued	Pump incorrectly installed	Check speed and direction of rotation
	Incorrect viscosity oil	Use only hydraulic oil of the correct viscosity
	Bad pump shaft seal	Pour oil around shaft at seal; if noise stops, seal is bad
Pump not delivering flow	Low oil level in reservoir	Add oil
	Pump air-locked	Prime pump
	Inlet air leak	Repair the leak
	Broken pump shaft or internal components	Check for damaged components, investigate the cause of failure
	Pump rotation reversed	Reverse rotation
Pump not delivering pressure	Pump not delivering flow	See above
	Relief valve setting incorrect	Check relief valve; if setting is correct inspect valve for worn or sticking components
	Oil leaking through component	Check through system to find
	Returning to reservoir	Leaking component usually will be hot
	Pump speed too low	Use tachometer, increase drive speed
	Pressure gauge defective	Replace gauge

Table 18-2   Directional Control Valve (DCV) Problems

Problem	Possible Cause	Cure
Valve sluggish	Dirt in system	Clean valve, clean fluid in system, change filters
	Internal drain blocked	Clear drain, clear line
	Mounting bolts too tight	Torque valve body bolts according to manufacturer's instructions
	Grounded solenoid coil	Check for ground, repair or replace coils
Valve fails to shift	Dirt in system	Clean valve, fluid, filters
	Blocked internal drain	Clean drain lines
	No pilot pressure	Check for source of pressure loss
	Solenoid voltage absent	Find location of voltage drop
	Mounting bolts too tight	Torque to correct specification
Valve actuation produces unusual response	Lines reversed	Check for proper connection
	Internal components installed incorrectly	Check prints to insure proper assembly

Table 18-3   Hydraulic Motor Problems

Problem	Possible Cause	Cure
Motor running in reverse direction	Lines crossed at motor Lines reversed at DCV	Reverse lines Reverse lines
Motor will not come up to speed, has no torque	Insufficient system pressure	Check system, relief valve, pump, etc.
External oil leakage	Internal motor leakage Drain line stopped up Seals and packing leaking	Repair or replace motor Clean line Repair or replace seals, packing
Motor will not turn over	Insufficient pressure for given load	Check load and system pressure

Table 18-4   Hydraulic Cylinder Problems

Problem	Possible Cause	Cure
Will not extend or retract	Insufficient flow or pressure	Check system for proper flow and pressure
	Too heavy load	Check load rating of equipment
	Mechanical bind	Check alignment of system, cylinder to load
	Internal leakage of cylinder	Repair or replace cylinder
		Can check by monitoring flow in return line, if there is flow, the cylinder is leaking through
Erratic action	Air in system	Bleed air from lines, find the leak and repair
	Internal leakage	Check to insure leakage, then repair or replace cylinder
	Cylinder binding	Misalignment, align cylinder
		Worn cylinder parts, repair or replace cylinder components

Chapter 19

Pneumatic Troubleshooting

As in hydraulic systems, pneumatic troubleshooting involves knowledge of system components. This knowledge will make the use of the following tables easier when troubleshooting pneumatic systems. Table 19-1 should be used as a guide for pneumatic troubleshooting.

Table 19-1   Pneumatic System Problems

Problem	Possible Cause	Cure
Low air pressure	Compressor volumetric output insufficient	Install a second compressor or a receiver in system
	Leaks in system	Repair all leaks
	Air filters are dirty	Clean or replace filters
	Internal compressor components worn	Repair or replace defective parts
Noisy operation	Defective components in compressor	Repair or replace defective part
	Inadequate lubrication	Increase lubrication
	Misalignment	Align correctly
Air temperature high between stages	Intercooler stopped up	Clean and flush intercooler
	Water temperature too hot	Cool water before entering intercooler
Insufficient air volume in system	Plugged inlet filter	Change or clean inlet filter
	Worn compressor component	Repair or replace worn parts
Oil in air lines	Worn compressor rings	Replace compressor rings
	Defective lubricator	Check flow rate of lubricator for proper amount
Water in system	Defective oil separator	Clean, drain, repair separator
	Inlet bringing in moist air	Reposition inlet
	Cooler not working	Repair coolers; drain, clean also to help with moisture removal
	Moisture separator not working	Service separator

Glossary

Abrasive A substance that wears or grinds away by friction.

Absolute A pressure scale where a vacuum is 0 psi and atmospheric pressure is 14.7 psi.

Acceleration The rate of change of an object's velocity.

Addendum The distance from the pitch line to the tip of the tooth in a gear.

Additive A substance added to a lubricant to improve or enhance a quality.

Alignment The proper positioning of two components in relation to one another.

Backlash The amount of clearance between two gear teeth in mesh.

Bevel A gear with teeth that are cut into the face of the gear. The gears may intersect at almost any angle.

Bleeding The process of an oil working out of the base in a grease.

Bolt A fastening device used with a nut to hold two or more parts together. The bolt is tightened by turning the nut.

Bushing A cylindrical device used to reduce the friction between two moving parts.

Cavitation A process in which dissolved air is removed from the fluid on the inlet side of the pump and implodes back into the fluid on the outlet side of the pump.

Chordal thickness   The thickness of a gear tooth measured on the pitch line. A straight line measurement.

Circular thickness   The thickness of a gear tooth measured on the pitch circle. This measurement is an arc. Usually only given on spur gears.

Clearance   The radial distance between the tip of a gear tooth and the bottom of the mating tooth space.

Compression Packing   Packing that accomplishes sealing by being deformed under pressure.

Corrosion   A process where material is worn away gradually, usually by some form of a chemical action.

Coupling   Device used to connect shafting.

Cylinder   A fluid power component that is used for linear motion. Sometimes called a linear actuator.

Deceleration   The process of reducing an object's velocity.

Dedendum   The distance from the pitch circle to the root of a tooth.

Density   The mass of a material for a given volume.

Depth of engagement   The radial engagement of a screw thread.

Dropping point   The temperature at which a grease liquifies.

Dynamic   A type of friction relating to objects in motion.

Efficiency   A ratio of the input energy to the output energy. Usually expressed as a percentage.

End-play   Motion along the axis of a shaft.

Fit   A designation used to indicate the closeness of two mating screw threads.

Flash point   The temperature at which a substance will burst into flames.

Friction   The resistance to motion of two bodies in contact. The three types are static, dynamic, and rolling.

Gauge   A pressure scale. This scale ignores atmospheric pressure. Atmospheric pressure is 0 psig.

Gear   A mechanical toothed wheel that provides a drive with a positive transmission of torque.

Helical   A form of gear tooth that's cut at a helix angle on the face of the gear.

Herringbone A gear with two sets of teeth with opposite hands cut into the face. The helix angle of both sets of teeth are the same.

Hub The center part of a coupling or sprocket.

Humidity The measure of the amount of water vapor in a given volume of air.

Hydrodynamic A physical property by which a rotating body can develop pressure in a surrounding fluid. Some bearings depend on this principle to support their load.

Hydrostatic A type of bearing that has fluid pressure supplied to it by an external source.

Hypoid bevel A bevel gear that has an offset axis. The shafting may be extended to provide more support for the gear.

Input Power or speed put into some mechanical or fluid power component.

Interference A form of wear in a gear drive where the two gears are in tooth tip to root contact.

Internal gear A gear with the teeth cut on the inside circumference of the pitch circle.

Keyway A groove or channel cut into a mechanical component for a key.

Lantern ring A ring in a stuffing box used to provide lubrication for the packing.

Lead The amount of axial distance traveled by the turning of the threaded component one turn.

Length of engagement The axial length that two threaded components are in contact.

Lubricant A substance introduced between two or more moving components to reduce friction and wear.

Major diameter The outside diameter of a screw thread, measured radially.

Mass The measure of the amount of material in an object.

Mesh The size of one of the openings in a filter or strainer.

Mesh The working contact of gear teeth.

Micron A unit of measure equal to .00039 inches.

Minor diameter The smallest diameter of a screw thread measured radially at the root of the thread.

Motor   Hydraulic device for converting flow to rotary motion. May also be a prime mover in a mechanical or fluid power system (electric motor).

Multiple threads   More than one set of threads progressing along the axial length of a threaded device.

Nominal   A term in rating a filter that refers to its approximate size in microns.

Nonpositive displacement   A condition in a pump or compressor that does not produce a given volume per revolution.

Number of threads   The number of threads per inch of axial length on a screw thread.

Orifice   A restriction to flow in a fluid power system.

Output   The power or speed delivered by a mechanical or fluid power drive.

Oxidation   The process of a material combining with oxygen. Usually results in the formation of rust.

Penetration   A test to check the thickness of a grease done by dropping a fixed weight (cone shaped) into the grease from a given height. The depth the cone penetrates is the grease's penetration rating.

Pin   A type of chain link used to connect two roller links.

Pinion   The smaller of two gears in a gear drive.

Pitch   The distance from a point on a gear to the corresponding point on the next tooth. In a chain drive, it's the distance from a point on one chain link to the corresponding point on the next link.

Pitch diameter   The diameter of an imaginary circle that connects all the pitch points on a gear or sprocket.

Planetary   A type of internal gear drive having several internal gears rotating around a center sun gear.

Pneumatics   A field of power transmission that uses a gas for the transmitting medium.

Positive displacement   A condition in a pump or compressor that displaces a certain volume for every revolution.

Power   A force moving through a distance in a given time period.

Pressure   Force per unit of area.

Pumpability   The ability of a lubricant to be pumped.

Rack and pinion   A type of gear drive that translates rotary motion to linear motion or vice versa.

Radial   A type of load on a bearing that is applied 90 degrees to the shaft axis.

Relative humidity   The ratio of the amount of water vapor that a given volume of air contains compared to the amount of water vapor that it could contain. Usually expressed as a percentage.

Roller   A type of chain link composed of a bushing, link plate, and a roller.

Root   In a gear drive, the bottom area between two teeth. In a screw thread, it's the lowest point between two threads.

Runout   A movement along the axis of an object, usually a shaft.

Screw   A simple machine that utilizes the principle of the inclined plane by wrapping the inclined plane around a cylinder.

Seal cage   Same as lantern ring; a device used to lubricate the packing in a stuffing box.

Spiral bevel   A bevel gear that has the teeth cut in a spiral form on the face of the gear.

Sprocket   A wheel that has teeth cut into its outside circumference for engagement with a chain.

Spur   A gear tooth form where the teeth are cut parallel to the shaft axis.

Static   A type of friction between two surfaces in direct contact but not in motion.

Stud   A fastening device threaded on each end.

Stuffing box   An assembly that uses packing to seal an area of low pressure from an area of high pressure.

Thrust   A type of load that's applied along a shaft axis.

Torque   A force applied at a distance that attempts to produce rotation. It doesn't have to produce motion to produce torque.

Velocity   The rate of motion in a given direction.

Viscosity   The measure of the fluid's resistance to flow.

Volume   The measure of the space an object occupies, measured in cubic units.

Weight   The measure of the gravitational attraction of the earth for an object.

Whole depth   In a gear drive, the radial measure of a tooth space.

Work   A force times the distance it moves. Motion must occur for work to take place.

Working depth   The sum of the addendums of two mating gears.

Worm   A type of gear drive for use at right angles. It uses a screw-shaped worm with a mating gear.

Zerol bevel   A bevel gear that uses spiral shaped teeth with a zero spiral angle.

Index